National Key Book Publishing Planning Project of the 13th Five-Year Plan

"十三五"国家重点图书出版规划项目

International Clinical Medicine Series Based on the Belt and Road Initiative

"一带一路"背景下国际化临床医学丛书

# Biochemistry Experiment Manual

## 生物化学实验指导

Chief Editor Li Ling Yu Hong

主编 李凌 喻红

郑州大学出版社

ZHENGZHOU UNIVERSITY PRESS

**图书在版编目(CIP)数据**

生物化学实验指导 = Biochemistry Experiment Manual：英文／李凌，
喻红主编． — 郑州：郑州大学出版社，2020.12
("一带一路"背景下国际化临床医学丛书)
ISBN 978-7-5645-7652-3

Ⅰ．①生…　Ⅱ．①李…②喻…　Ⅲ．①生物化学 - 化学实验 -
医学院校 - 教学参考资料 - 英文　Ⅳ．①Q5-33

中国版本图书馆 CIP 数据核字(2020)第 257795 号

**生物化学实验指导 = Biochemistry Experiment Manual：英文**

| | | | | |
|---|---|---|---|---|
| 项目负责人 | 孙保营　杨秦予 | | 策 划 编 辑 | 李龙传 |
| 责 任 编 辑 | 陈文静　张 楠 | | 装 帧 设 计 | 苏永生 |
| 责 任 校 对 | 薛 晗 | | 责 任 监 制 | 凌 青　李瑞卿 |

| | | | | |
|---|---|---|---|---|
| 出版发行 | 郑州大学出版社有限公司 | | 地　　址 | 郑州市大学路 40 号(450052) |
| 出 版 人 | 孙保营 | | 网　　址 | http://www.zzup.cn |
| 经　　销 | 全国新华书店 | | 发行电话 | 0371-66966070 |
| 印　　刷 | 河南文华印务有限公司 | | | |
| 开　　本 | 850 mm×1 168 mm　1 / 16 | | | |
| 印　　张 | 18.25 | | 字　　数 | 700 千字 |
| 版　　次 | 2020 年 12 月第 1 版 | | 印　　次 | 2020 年 12 月第 1 次印刷 |

| | | | | |
|---|---|---|---|---|
| 书　　号 | ISBN 978-7-5645-7652-3 | | 定　　价 | 69.00 元 |

# Staff of Expert Steering Committee

**Chairmen**

Zhong Shizhen    Li Sijin    Lü Chuanzhu

**Vice Chairmen**

Bai Yuting       Chen Xu       Cui Wen          Huang Gang      Huang Yuanhua
Jiang Zhisheng   Li Yumin      Liu Zhangsuo     Luo Baojun      Lü Yi
Tang Shiying

**Committee Member**

An Dongping      Bai Xiaochun     Cao Shanying    Chen Jun        Chen Yijiu
Chen Zhesheng    Chen Zhihong     Chen Zhiqiao    Ding Yueming    Du Hua
Duan Zhongping   Guan Chengnong   Huang Xufeng    Jian Jie        Jiang Yaochuan
Jiao Xiaomin     Li Cairui        Li Guoxin       Li Guoming      Li Jiabin
Li Ling          Li Zhijie        Liu Hongmin     Liu Huifan      Liu Kangdong
Song Weiqun      Tang Chunzhi     Wang Huamin     Wang Huixin     Wang Jiahong
Wang Jiangang    Wang Wenjun      Wang Yuan       Wei Jia         Wen Xiaojun
Wu Jun           Wu Weidong       Wu Xuedong      Xie Xieju       Xue Qing
Yan Wenhai       Yan Xinming      Yang Donghua    Yu Feng         Yu Xiyong
Zhang Lirong     Zhang Mao        Zhang Ming      Zhang Yu'an     Zhang Junjian
Zhao Song        Zhao Yumin       Zheng Weiyang   Zhu Lin

# 专家指导委员会

# Staff of Editor Steering Committee

**Chairmen**

Cao Xuetao    Liang Guiyou    Wu Jiliang

**Vice Chairmen**

Chen Pingyan    Chen Yuguo    Huang Wenhua    Li Yaming    Wang Heng

Xu Zuojun    Yao Ke    Yao Libo    Yu Xuezhong    Zhao Xiaodong

**Committee Member**

| | | | | |
|---|---|---|---|---|
| Cao Hong | Chen Guangjie | Chen Kuisheng | Chen Xiaolan | Dong Hongmei |
| Du Jian | Du Ying | Fei Xiaowen | Gao Jianbo | Gao Yu |
| Guan Ying | Guo Xiuhua | Han Liping | Han Xingmin | He Fanggang |
| He Wei | Huang Yan | Huang Yong | Jiang Haishan | Jin Chengyun |
| Jin Qing | Jin Runming | Li Lin | Li Ling | Li Mincai |
| Li Naichang | Li Qiuming | Li Wei | Li Xiaodan | Li Youhui |
| Liang Li | Lin Jun | Liu Fen | Liu Hong | Liu Hui |
| Lu Jing | Lü Bin | Lü Quanjun | Ma Qingyong | Ma Wang |
| Mei Wuxuan | Nie Dongfeng | Peng Biwen | Peng Hongjuan | Qiu Xinguang |
| Song Chuanjun | Tan Dongfeng | Tu Jiancheng | Wang Lin | Wang Huijun |
| Wang Peng | Wang Rongfu | Wang Shusen | Wang Chongjian | Xia Chaoming |
| Xiao Zheman | Xie Xiaodong | Xu Falin | Xu Xia | Xu Jitian |
| Xue Fuzhong | Yang Aimin | Yang Xuesong | Yi Lan | Yin Kai |
| Yu Zujiang | Yu Hong | Yue Baohong | Zeng Qingbing | Zhang Hui |
| Zhang Lin | Zhang Lu | Zhang Yanru | Zhao Dong | Zhao Hongshan |
| Zhao Wen | Zheng Yanfang | Zhou Huaiyu | Zhu Changju | Zhu Lifang |

# 编审委员会

# Editorial Staff

| | |
|---|---|
| Wang Huilian | Xi'an Jiaotong University Health Science Center |
| Wang Kai | Hainan Medical University |
| Wang Xuejun | Nanjing Medical University |
| Wang Yanfei | Guangzhou Medical University |
| Wu Ning | Guizhou Medical University |
| Wu Yanhui | Jinan University School of Medicine |
| Xu Yan | School of Basic Medical Sciences, Zhengzhou University |
| Yang Guang | College of Basic Medical Sciences, Jilin University |
| Yin Hong | Southern Medical University |
| Yin Ye | Nanjing Medical University |
| Yu Hailang | Southern Medical University |
| Yu Hong | Wuhan University School of Basic Medical Sciences |
| Yuan Ping | Tongji Medical College, Huazhong University of Sciences and Technology |
| Zhang Chunjing | Qiqihar Medical University |
| Zhang Weijuan | School of Basic Medical Sciences, Henan University |
| Zhang Xu | School of Basic Medical Sciences, Zhengzhou University |
| Zhang Yuzhe | College of Basic Medical Sciences, Dali University |
| Zhou Hongbo | Harbin Medical University |
| Zhou Ti | Zhongshan School of Medicine, Sun Yat-sen University |

**Secretary**

| | |
|---|---|
| Du Fen | Wuhan University School of Basic Medical Sciences |

# 作者名单

主审
　　药立波　中国人民解放军空军军医大学
主　编
　　李　凌　南方医科大学
　　喻　红　武汉大学基础医学院
副主编
　　吕立夏　同济大学医学院
　　周宏博　哈尔滨医科大学
　　都　建　安徽医科大学
　　林　利　兰州大学基础医学院
　　胡晓鹃　南昌大学基础医学院
编　委（以姓氏汉语拼音排序）
　　蔡翠霞　南方医科大学
　　杜　芬　武汉大学基础医学院
　　都　建　安徽医科大学
　　范新炯　安徽医科大学
　　费小雯　海南医学院
　　胡晓鹃　南昌大学基础医学院
　　李　姣　同济大学医学院
　　李　凌　南方医科大学
　　李志红　三峡大学医学院
　　林贯川　南方医科大学
　　林　利　兰州大学基础医学院
　　刘安玲　南方医科大学
　　刘宝琴　中国医科大学
　　吕立夏　同济大学医学院
　　彭　帆　三峡大学医学院
　　史　磊　新乡医学院
　　孙　巍　吉林大学基础医学院
　　王华芹　中国医科大学

王慧莲　西安交通大学医学部
王　凯　海南医学院
王学军　南京医科大学
王燕菲　广州医科大学
吴　宁　贵州医科大学
吴颜晖　暨南大学医学院
徐　琰　郑州大学基础医学院
杨　光　吉林大学基础医学院
尹　虹　南方医科大学
尹　业　南京医科大学
余海浪　南方医科大学
喻　红　武汉大学基础医学院
袁　萍　华中科技大学同济医学院
张春晶　齐齐哈尔医学院
张维娟　河南大学基础医学院
张　旭　郑州大学基础医学院
张钰哲　大理大学基础医学院
周宏博　哈尔滨医科大学
周　倜　中山大学中山医学院

秘　书

杜　芬　武汉大学基础医学院

# Preface

At the Second Belt and Road Summit Forum on International Cooperation in 2019 and the Seventy-third World Health Assembly in 2020, General Secretary Xi Jinping stated the importance for promoting the construction of the "Belt and Road" and jointly build a community for human health. Countries and regions along the "Belt and Road" have a large number of overseas Chinese communities, and shared close geographic proximity, similarities in culture, disease profiles and medical habits. They also shared a profound mass base with ample space for cooperation and exchange in Clinical Medicine. The publication of the International Clinical Medicine series for clinical researchers, medical teachers and students in countries along the "Belt and Road" is a concrete measure to promote the exchange of Chinese and foreign medical science and technology with mutual appreciation and reciprocity.

Zhengzhou University Press coordinated more than 600 medical experts from over 160 renowned medical research institutes, medical schools and clinical hospitals across China. It produced this set of medical tools in English to serve the needs for the construction of the "Belt and Road". It comprehensively coversaspects in the theoretical framework and clinical practicesin Clinical Medicine, including basic science, multiple clinical specialities and social medicine. It reflects the latest academic and technological developments, and the international frontiers of academic advancements in Clinical Medicine. It shared with the world China's latest diagnosis and therapeutic approaches, clinical techniques, and experiences in prescription and medication. It has an important role in disseminating contemporary Chinese medical science and technology innovations, demonstrating the achievements of modern China's economic and social development, and promoting the unique charm of Chinese culture to the world.

The series is the first set of medical tools written in English by Chinese medical experts to serve the needs of the "Belt and Road" construction. It systematically and comprehensively reflects the Chinese characteristics in Clinical Medicine. Also, it presents a landmark achievement in the implementation of the "Belt and Road" initiative in promoting exchanges in medical science and technology. This series is theoretical in nature, with each volume built on the mainlines in traditional disciplines but at the same time introducing contemporary theories that guide clinical practices, diagnosis and treatment methods, echoing the latest research findings in Clinical Medicine.

As the disciplines in Clinical Medicine rapidly advances, different views on knowledge, inclusiveness, and medical ethics may arise. We hope this work will facilitate the exchange of ideas, build common ground while allowing differences, and contribute to the building of a community for human health in a broad spectrum of disciplines and research focuses.

*Nick Lemoine*

**Foreign Academician of the Chinese Academy of Engineering**

**Dean, Academy of Medical Sciences of Zhengzhou University**

**Director, Barts Cancer Institute, London, UK**

**6th August, 2020**

# Foreword

Biochemistry is the chemistry of life. Theories and techniques of biochemistry and molecular biology have been widely used in various fields of medicine and life science. Therefore, laboratory techniques of biochemistry have become an important tool to reveal the mysteries of life. Here we organized 37 young and middle-aged backbone teachers from 24 medical universities to compile this experimental textbook.

This manual includes 6 parts and an appendix. The content can be divided into three portions: general introduction to biochemistry experiments (Part I), special experiments (Part II–VI), and the appendix. ① The general introduction includes basic requirements and basic skills of biochemistry experiments (Chapter 1–2), and the basic principles of four biochemistry techniques (Chapter 3–6). ②Special experiments are laid out according to the system, ranging from protein experiments, enzymology experiments, carbohydrate and lipid experiments, to nucleic acid experiments (Chapter 7–21). ③ The basic procedure, basic route and methods of designing experiments and bioinformatics experiments are described in Chapter 22–24. ④The appendix includes commonly used data in biochemistry experiments.

This textbook has several distinguishing features. ①The content is systematic, ranging from basic requirements and skills of experiments to special experiments. ②The content is rich and detailed. The total of 35 experiments includes basic experiments, comprehensive experiments, designing experiments, and researching experiments. ③Writing features: in each experiment, after the principle is briefly introduced, the key points, technique analysis and notes related to the procedure are intensively described. Finally, the problems that may be encountered, as well as potential causes and solutions are analyzed. Therefore, this book is not only a useful experiment textbook but also an important reference manual for laboratory techniques related to biochemistry and molecular biology.

The intended audience for this book includes Bachelor of Medicine and Bachelor of Surgery(MBBS) students, extended-year medical students, and medical school graduates. This manual is also suitable for experimental technicians of biochemistry and molecular biology, in addition to students, teachers and scientific researchers with related majors. Applicable courses include those related to biochemistry experiments (undergraduate) and molecular biology experiments (undergraduate and graduate students). Each school may offer corresponding experiment content according to different professional requirements and laboratory conditions.

The participating universities have provided extensive support for the compilation of this book. We are very grateful to all experts and all universities.

Due to the editors' limited level of expertise, this book inevitably has omissions and inaccurate aspects, and we really appreciate your feedback so we can improve this manual.

*Authors*

# Brief Introduction

The content of this manual is divided into three portions: general introduction to biochemistry experiment (Part I), special experiments (Part II – VI), and appendix. The general introduction includes basic requirements and basic skills of biochemistry experiments, the basic principles of biochemistry techniques (Chapter 1–6). Special experiments are laid out according to the system which range from protein experiments, enzymology experiments, carbohydrate and lipid experiments, nucleic acid experiments, to experimental design and bioinformatics experiments. This textbook is systematic and detailed, includes a total of 37 experiments. In each experiment, after the principle is briefly introduced, the key points, technique analysis and notes of the procedure are intensively described. Finally, the problems that may be encountered, as well as potential causes and solutions are analyzed. This manual is not only a useful experiment textbook but also an important reference book of laboratory techniques for biochemistry and molecular biology.

# Content

## Part I  General Introduction to Biochemistry Experiment

# Part Ⅱ   Protein Experiments

# Part Ⅲ  Enzymology Experiments

# Part Ⅳ  Carbohydrate and Lipid Experiments

# Part V　Nucleic Acid Experiments

# Part Ⅵ　Design of Experiment and Informational Experiment

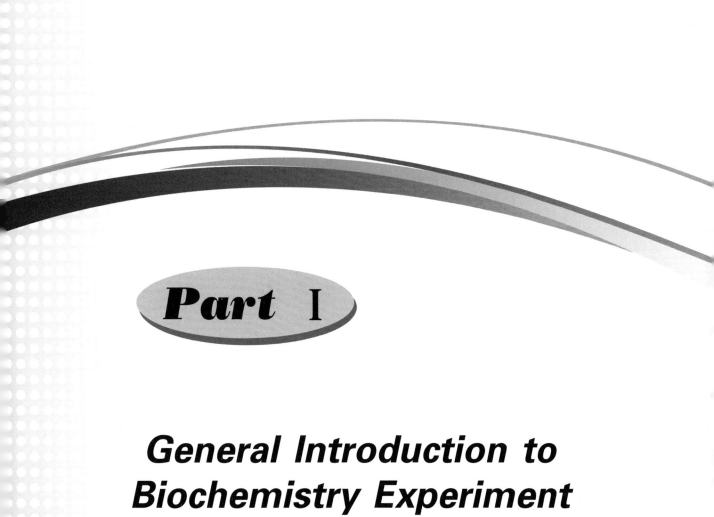

# Part I

# General Introduction to Biochemistry Experiment

# Chapter 1

# Basic Requirements of Biochemistry Experiment

Bioactive substances are the subjects of biochemical experiments. In the experiment, these substances are easily affected by external environmental factors, resulting in differences in experimental results. Therefore, it is very important to comply with basic laboratory requirements as well as to carefully and realistically complete the experiment record and data processing and analysis.

## Section 1    Basic Laboratory Requirements

## 1.1    Biochemistry laboratory rules

### 1.1.1    Preview before class

Carefully preview the contents of the experiment before class. For example, become familiar with the experimental objective, principle, and procedure, understand how to use the instruments, and write a preview report.

### 1.1.2    Observe classroom discipline

Wear a white lab coat in the laboratory, do not be late or leave early. Keep quiet and do not eat in the laboratory.

### 1.1.3    Follow operating instructions

Carry out the experiment according to the experimental procedure. During the experiment process, do not deal with problems that cannot be solved by yourself without prior preparation. Consult the instructor when appropriate.

(1) Use the instruments in accordance with operating procedures. If instruments fail or are damaged, please turn them off immediately and report it to the instructors. Do not try to fix them by yourself.

(2) Cover the reagent immediately after use and put it back in its original place to avoid contamination.

(3) It is prohibited to discard solid trash such as waste papers into the sink.

### 1.1.4　Truthfully record and analyze

Objectively record the experimental results and carefully analyze them. If the experiment fails, you must carefully look for the reason and cannot arbitrarily modify the result. After the experiment, carefully write the lab report and submit it on time.

### 1.1.5　Abide by the principle of economy

Conserve drugs, reagents, articles, water, and electricity and take good care of public property. You are required to pay proper compensation for losses resulting from your own actions.

### 1.1.6　Keep safe

Pay attention to the safe use of water, electricity, and reagents. Stay away from fire when using inflammable and explosive materials. When using a test tube to heat, make sure the nozzle is not facing others. Prevent strong acid, alkali, and toxic substances from being sucked into your mouth or splashed on others. Do not throw strong acid, strong alkali, high temperature, and toxic substances on the test bench and instruments.

### 1.1.7　Keep clean

The test bench should be kept clean and tidy, and instruments and medicine should be arranged neatly. After the experiment is completed, the experimental equipment shall be washed and put away, the experimental table shall be wiped clean, and the instruments used shall be closed and restored to their original state. On a daily basis, experimenters must be responsible for cleaning and tidying up the entire laboratory and keeping it tidy and clean. Check the safety of power supplies, water sources, doors, and windows before leaving the laboratory, and strictly enforce the daily registration system. You are allowed to leave after receiving the teacher's approval.

## 1.2　Biochemistry laboratory safety

Fragile glass instruments, sophisticated analytical instruments, and corrosive, toxic and flammable chemicals are often used in laboratories, and these present potential hazards of explosions, fires, poisoning, burns, cuts, electric shocks, and other accidents. Therefore, it is necessary to strictly abide by laboratory rules and experiment procedures, as well as learn certain knowledge about safety self-help and accident handling. If an accident occurs, keep calm, immediately report it to the teacher and deal with it according to the specific situation.

(1) Do not eat anything in the laboratory. All chemicals cannot enter the mouth, laboratory vessels cannot be used to hold food, and food cannot be stored in the laboratory's refrigerator. After the experiment is finished, wash your hands and leave the lab.

(2) It is necessary to carefully study experimental procedures and relevant safety technical regulations, understand the equipment's performance and possible causes of accidents during operation as well as master methods for preventing and handling accidents.

(3) Be careful when using concentrated acids, concentrated alkalis, and chromic acid lotions, and do not splash them into your eyes, skin, or clothing. If solutions accidentally splash on the skin and eyes, immediately rinse them with a large amount of tap water.

(4) If toxic drugs are swallowed by mistake, spit them out immediately and gargle repeatedly with fresh water. Those who are seriously injured should be sent to the hospital immediately after emergency treatment.

(5) Do not touch the power supply with wet hands. If electric shock occurs, the power should be cut off

quickly and artificial respiration should be carried out if necessary. In case of scalding, rinse immediately with running water or immerse the affected area in cold water. Wounds should be simply bandaged and wounded personnel should be sent to the hospital for treatment. Treatment for severe burns should not be carried out on site, and you should seek immediate treatment at a nearby hospital.

(6) Carefully use delicate glass instruments. If you are accidentally cut by glass, you should check the wound for glass fragments. After picking out the fragments, minor injuries can be coated with mercurochrome, violet or iodine, and then bandaged. When the wound is severe, it should be treated in a clinic or hospital as soon as possible after simple treatment.

(7) If a fire occurs during an experiment, the power or gas source should be immediately cut off, and a suitable fire extinguishing method should be chosen quickly to identify the cause of the fire. If the fire is caused by alcohol, benzene or ethyl ether and is relatively small, the fire can be covered by a damp cloth, asbestos cloth or sand. A foam extinguisher can be used when the fire is large. In the event of a fire in an electrical appliance, it is necessary to switch off the power supply before using carbon dioxide and carbon tetrachloride fire extinguishers. While extinguishing the fire, it is necessary to quickly remove flammable and explosive materials to prevent spreading the fire. Do not panic when your clothes are on fire. You should take off your clothes as soon as possible, cover the fire with asbestos cloth, or lie down and roll on the spot. Immediately call the police in case of emergency.

# Section 2    Experimental Record and Data Processing

## 2.1    Experimental record

During the experiment process, in order to obtain accurate experimental results, it is necessary not only to correctly observe and measure but also to correctly record observed experimental phenomena, conditions, and data. When recording and expressing data results, they should not only reflect the size of measured values but also reflect their accuracy. Significant figures can reflect the measured values' credibility. Correctly using significant figures and their calculation roles is one of the basic skills of experimental technicians.

### 2.1.1    The contents of the experimental record

The experimental record is one of the important links in experimental teaching and scientific research. Correct recording of experimental data is the basic requirement for cultivating a rigorous scientific style. The experimental record should be true, complete, standardized and clear.

The following three main contents should be recorded in detail in the experiment. ① Experimental conditions: for example, the material's source and quality, the reagent's manufacturer, specification, dosage and concentration, and the experiment time, skill and error in key points of operation, in order to check and find the reasons for success or failure as a reference. ② Phenomena: such as the solution's color change after adding the reagent, etc. ③ Raw data: design the experimental data table (note the three-line tabular format) to accurately record raw data measured in the experiment. When recording measurement data, the number of significant digits should be correctly recorded.

The experimental data and phenomena should be immediately and accurately recorded. Experiment records should be recorded with a pen or ballpoint pen. Pencil should not be used. The records must be objective and truthful and must not be fabricated or falsified. If an error is found, the data should be crossed out and the correct number written above it.

## 2.1.2　Significant figures

The significant figures of a number are those digits contributing to its precision, including all accurate-figures and the last estimate figure. They are closely related to measurement precision and can't be increased or decreased arbitrarily. When using significant figures in biochemical experiments, the following points should be noted.

(1) Correctly record the measurement data: the recorded data must faithfully reflect the actual measurement's accuracy.

(2) Determine the correct sample size and select the appropriate instrument: mass analysis or volumetric analysis are commonly used for determining constant components. The method's accuracy can reach 0.1%. Therefore, the error of each step in the entire measurement process should be less than 0.1%. When weighing samples with the analytical balance, the sample size should be greater than 0.2 g in general so that the weighing error is less than 0.1%.

(3) Correctly report the analysis results: the analysis results's accuracy should faithfully reflect that of each measurement step and should not be higher than that of the step with the largest error in each measurement step.

(4) Correctly understand the accuracy requirements: the error in quantitative analysis of biochemical experiments is objective, and the accuracy requirement depends on the need and objective possibility. Common gravimetric and volumetric methods are used to determine constant components, the method error is about ±0.1%, and generally four significant figures are used. For the analysis of trace substances, the relative error of the analysis results can be within ±2% –±3%, which meets actual needs.

(5) The trade−off of significant digits in a calculator result: the use of electronic calculators has become very popular, and makes it very convenient to calculate the number of digits. However, do not copy the figures shown on the calculator when recording the results. Determine the number of digits in calculator results according to the rules of significant figures.

## 2.2　Data processing

Appropriate processing methods should be used for sorting out and analyzing a series of values obtained in the experiment, the quantitative relationship of the subjects studied can be accurately reflected. In biochemical experiments, the tabulation method or graphic method is usually used to express the experimental results, so as to make the results clear, and may also reduce and make up for some measurement errors. According to a series of measurements of standard samples, a table or a standard curve can also be drawn, and the results can be directly detected by the measured values.

(1) Tabulation method: list the data in the appropriate tables. The data should correctly reflect the significant figures of measurement, and the error value should be calculated if necessary.

(2) Graphic method: the graphic method is a method to represent data and obtain analysis results for drawing pictures. In other words, the experimental data are drawn in graphs according to the corresponding relationship between the independent variable and the dependent variable to obtain the required analysis results. This method is widely used in instrumental analysis, such as when using standard curve method to calculate sample concentration, or using absorption curve in spectrophotometry to determine spectral characteristics data and qualitative and quantitative analysis.

(3) Representation of mathematical equations: the method to express the relationship between variables using mathematical equations is called mathematical equation representation, also known as the analytic method. A large number of experimental data are summarized and processed, from which the functional relations of various physical quantities are summarized. This expression is concise and accurate, and can quickly

calculate relevant results such as solution concentration, interpolation, differentiation, integration, etc. The most commonly used analytical method is the regression equation method, which calculates the regression equation using the regression analysis of the data pairs of two variables and then calculates the number of components that the variables will measure.

# 2.3　Requirements of the laboratory report

By analyzing and summarizing the experiment's results and problems, the lab report deepens understanding and mastery of relevant theories and techniques, and improves the ability to analyze, synthesize and summarize problems. It is also a process of learning how to write research papers. The basic requirements for contents of lab reports are as follows.

(1) Principle: the experiment's principle is briefly described and expressed by chemical reaction equations when chemical reactions are involved.

(2) Materials: biological samples from various sources and reagents and major instruments should be included. Avoid using uncommonly accepted trade names and generic names for chemical agents. The concentration of reagents should be clearly labeled.

(3) Procedures: the description should be concise and cannot be copied from lab protocol. It can be represented by using a process flow chart or self-designed table, but the key links of operation and experimental conditions should be written clearly so that others can repeat the experiment.

(4) Results: the experiment results (such as observed phenomena) and data should be sorted, summarized, analyzed and compared, and the experiment results should be summarized as charts, such as the comparison table between the experimental group and the control group, etc.

(5) Discussion: the discussion is not a restatement of experimental results, but a logical deduction based on the results. For qualitative experiments, reasonable conclusions based on analyzing the experimental results should be drawn. It may also include some questions about experimental methods, operating techniques and related experiments, analysis, and comments on abnormal experimental results, in addition to understanding, experience, and suggestions on experimental design, ideas for improvement of experimental courses, etc.

(6) Conclusion: generally, the experiment should have a conclusion that should be simple and concise while also explaining the result obtained in this experiment.

*Yin Hong, Liu Anling*

# Chapter 2

## Basic Skills in Biochemistry Experiment

Biochemical experiments involve many basic operations, such as cleaning of glass equipment, mixing of solutions, centrifugation, and the use of pipette and micropipette, etc. If these operations are not standardized, this will affect the accuracy of experimental results. Therefore, it is very important to master the basic operation skills of biochemistry experiment for obtaining accurate experimental results.

## Section 1　Mixing of Solutions

When preparing a solution or reaction system, it must be thoroughly mixed. Mixing methods vary, depending on the size and shape of the container and the amount and nature of the solution in the container.

## 1.1　Vibration mixing

This operation only depends on the mechanical force generated by the operator without the aid of tools. There are several specific operation methods.

(1) Sway mixing: hold the upper part of the test tube with your hand and gently shake to mix the liquid. It is suitable for the test tube when there is less liquid.

(2) Flick mixing: hold the upper end of the container with one hand, flick or move the bottom of the container with the other hand's finger to make the liquid swirl in the container. It is suitable for mixing conical tubes, small tubes, and Eppendorf tubes, etc.

(3) Whirling mixing: hold the top end of the container, and whirl the bottom with your wrist, elbow or shoulder as an axis. Do not vibrate up and down. It is suitable for mixing conical flasks, test tubes and small mouth containers without full solutions.

(4) Rotation mixing: hold the top of the container so that its bottom moves in a rapid circle on the table. It is suitable for mixing viscous solutions, but when the amount of liquid is not too full, it is appropriate to account for a container volume of 1/3-2/3.

(5) Inverted mixing: it is suitable for mixing contents of containers with a plug, such as a volumetric flask. The operation is intended to repeatedly invert the container.

(6) Pour mixing: it is suitable for mixing solutions in containers with a large liquid quantity and small inner diameter. Using two clean containers, pour slowly into the container along the wall and pour back and forth several times to achieve the purpose of mixing.

## 1.2   Mixing solutions with instruments

(1) Glass rod mixing: it is suitable for mixing beaker contents, such as dissolution and mixing of solid reagents. The thickness and length of the glass rod must be proportional to the size of the container and the amount of solution prepared. For example, do not use a long and thick glass rod to stir a small amount of solution in a small centrifuge tube. When stirring, try to make the glass rod move along the wall of the tube without stirring into air or splashing the solution.

(2) Pipette mixing: it is suitable for the dilution of samples with different concentration levels. Draw the solution with a pipette, and then forcefully blow the liquid back into the solution. Repeatedly suck and blow several times to make the solution fully mixed.

(3) Micropipette mixing: when adding the trace reagent to the solution, place the tip of micropipettes lightly under the liquid surface, and blow repeatedly to make it fully mixed. The tip should not be used again due to the contaminated solution.

(4) Grinding and mixing: when preparing the colloidal solutions, make the pestle move along with the direction of the mortar, do not grind back and forth.

(5) Shaker mixing: use the oscillator to shake the contents of the container to achieve the purpose of mixing.

(6) Magnetic stirrer mixing: it is suitable for automatic acid-base titration, pH gradient titration, etc. Put the beaker containing the solution to be mixed on the magnetic stirrer, and place a stirrer enclosed in a glass or plastic tube in the beaker, use electromagnetic force to make the stirrer rotate, so as to achieve the purpose of mixing the solution in the beaker.

## Section 2   The Use of Graduated Pipettes

## 2.1   Types of the graduated pipettes

Two types of graduated pipettes are commonly used in biochemistry. One is full outflow. Its scale is marked to the tip of the pipette, the capacity includes all liquid, and residual liquid at the tip should be blown out when the liquid is released. The graduated pipette is marked "blow" at the top of the tube. The other is partial outflow. When the liquid is released, it naturally flows out. The tip remains in the tube for a few seconds, and the remaining liquid must not be blown out. The graduated pipette is marked "fast" on the top of the tube, indicating that the pipette has corrected the error of the tip's residual liquid, so it cannot blow out the residual liquid. Therefore, before using the pipette, it is important to determine whether it has the word "blow".

## 2.2   The procedure of using graduated pipettes

(1) Choose a suitable graduated pipette: choose a suitable pipette according to your needs, and its capacity is better if it is equal to or slightly larger than the amount of liquid.

(2) Take the tube: hold the upper end of the pipette with the thumb and middle finger of your right

hand, and control the flow of liquid by blocking the top of the pipette with the index finger. The ring finger and pinky finger are naturally attached to the pipette, and keep the straw in a vertically stable state.

(3) Suck the liquid: pinch and press the ear washing bulb with the left hand, and place the pipette into the reagent. Aim the outlet of the lower end of the ear ball at the upper mouth of the pipette, gently suck the liquid up to 1−2 cm above the desired scale, quickly press the upper mouth of the tube with the index finger, so that the liquid will not flow out from the lower mouth of the tube pipette.

(4) Calibrate the scale: after removing the pipette from the solution, if there is a viscous liquid such as serum, clean the liquid on the outer wall of the tube tip with filter paper, and then use the index finger to control the liquid to descend to the desired scale. When observing the scale, the operator's line of sight, liquid concave, and the scale line should be on the same horizontal plane.

(5) Release the liquid: transfer the pipette with liquid into the new container, put the pipette tip on the inner wall of the container, but do not insert into the original liquid of the container. Release the index finger of your right hand and let the solution flow naturally with the container being tilted 15°−20°. The residual liquid is blown out or not according to the requirements of the graduated pipette.

(6) Washing: rinse the pipette immediately with tap water to prevent dry blockage if taking blood, serum and other viscous liquid samples (such as urine). The pipette that absorbs the general reagent does not need to be rinsed immediately; rinse it with tap water after the experiment is finished. Let it dry and soak in the chromic acid solution for several hours, then rinse it with tap water and distilled water (Figure 2.1).

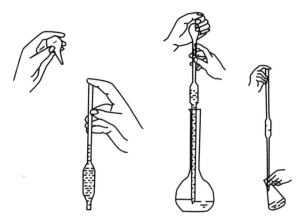

Figure 2.1　**The use of a graduated pipette**

## Section 3　The Use of Adjustable Micropipettes

Adjustable micropipette, also known as a sample gun, is a precise liquid sampling instrument with continuously adjustable liquid volume. Its range generally includes 10 μL, 20 μL, 100 μL, 200 μL, up to 1,000 μL. The suction head, also known as the suction nozzle, is commonly known as the pipette tip. The micropipette's working principle is to discharge air by pressing the moving mandrel, placing the suction head into the liquid, relaxing pressure on the button, using the built-in spring's mechanical force, and the button recovering to form negative pressure and suck the liquid. Micropipettes are precision instruments and must be handled and stored with care so as not to affect their accuracy. Choose a micropipette with a suitable volume.

# 3.1　The procedure of using adjustable micropipette

(1) Select a suitable micropipette: select a suitable micropipette according to your needs, and its capacity should be equal to or slightly greater than the fluid volume.

When sucking a dilute solution, use the positive pipetting method as follows (Figure 2.2).

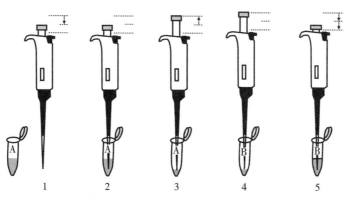

Figure 2.2　The positive pipetting method
1. Press the take button to first gear. 2. Keep the micropipette vertical for as far as possible and immerse the tip into the solution. 3. Release the button slowly to suck the solution. 4. Move the micropipette to the specified container B. 5. Slowly press the button to the second gear and completely release the solution.

(2) Adjust to the required value: determine the direction of the adjustment wheel before adjusting the button (adjustment wheel), and then adjust it to one grid greater than or less than the desired value, and finally adjust it back to the desired value.

(3) Loading tip: put the tip on the sucker rod and rotate gently to strengthen the seal.

(4) Take liquid: hold the micropipette vertically and press the button to the first gearwith your thumb, and immerse the tip a few millimeters below the liquid level, then slowly release the button, wait for 1 – 2 seconds and remove from the liquid.

(5) Release the liquid: remove the micropipette to the sample container, slowly press the button to the first gear, and wait for 1 –2 seconds, then press the button down completely (to the second gear) to release all the liquid. The tip should be taken out by sliding it upward along the container wall, and then the button should be released to make it reset, which completes the process. If the used tip does not work, press the other button on the top of the gun with your thumb and place the tip into the waste container.

(6) The reverse pipetting method: when sucking viscous or foamed liquids, the reverse pipetting method is required to prevent bubbles or foam from resulting in inaccurate fluid volume.

Press the button to the second gear, suck the liquid, and then release the button to recover. When releasing the liquid, press the button only to the first gear, and the residual liquid can be discarded together with the tip (Figure 2.3).

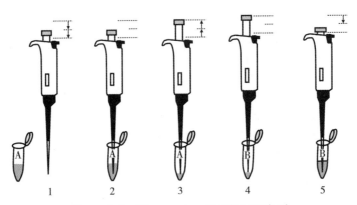

Figure 2.3    The reverse pipetting method

1. Press the take button to second gear. 2. Keep the micropipette vertical as far as possible and immerse the tip into the solution. 3. Release button slowly to suck solution. 4. Move the micropipette to the specified container B. 5. Slowly press the button to the first gear and release the required volume solution.

## 3.2    Notes

(1) The pipette is a precision instrument, and the adjustment wheel should be adjusted before taking the liquid.

(2) In the positive pipetting method, push the button to the second gear to clear the liquid while releasing the liquid.

(3) Inate errors.

(4) Do not move the button too fast. It will create bubbles in the tip and affect the amount of suction liquid.

(5) The shifter cannot be laid flat and inverted when a solvent is in the micropipette.

(6) The micropipette should be restored to its maximum capacity after use.

*Cai Cuixia, Liu Anling*

# Chapter 3

# Spectrophotometry

Given its' specific molecular structure, every compound has its own specific absorption spectrum. Spectrophotometry (absorption spectrometry) is widely used for both qualitative and quantitative analysis of compound via its specific absorption spectrum. Based on the wavelength which spectrophotometry deals with, it is divided into ultra–violet, visible and near–infrared regions. In biochemical experiments, spectrophotometry is mainly applied in quantification and qualification of amino acids, proteins, nucleic acids, and in the study of enzyme kinetics.

## Section 1   Basic Principles

When a beam of natural light passes through a solution containing colored compounds, certain ranges (wavelengths) of light are selectively absorbed (Figure 3.1). Each kind of compound has an individual set of energy levels associated with the composition of its chemical bonds and nuclei, and thus absorbs light at a specific wavelength. Therefore the plots of the absorption spectrum can be used to characterize compounds. Photometry involves both qualitative and quantitative usages of light absorption data obtained from samples.

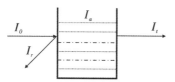

Figure 3.1   When a beam of light passes through a solution

$I_0$: the initial light intensity. $I_a$: the absorbed light intensity. $I_r$: the reflected light intensity. $I_t$: the transmitted light intensity.

Transmittance (T) means the ratio of the transmitted light intensity over the initial light intensity.

$$T = I_t / I_0 \times 100\%$$

The light absorption is described as absorbance (A), also could be called as optical density (OD) or degree of extinction (E).

$$A = -\lg T = -\lg I_t / I_0 = \lg I_0 / I_t$$

## 1.1    Lambert–Beer's law

If a beam of monochromatic light passed through a solution, part of the light is absorbed by the solution. The absorption of a given solution is proportional to both the length of the light pathway in the sample solution ( "$L$") and its concentration ( "$C$") .

$$A = K \times L \times C$$

Where $K$ is termed as the extinction coefficient ( E) , representing the ability of light absorption of solute. Note that $K$ value differs along with the changes in the wavelength of light and the solvent.

## 1.2    The application of Lambert–Beer's law

From the equation of Lambert–Beer's law, it is easy to deduce that the absorbance of a solution is proportional to its concentration, assuming the length of the light path and the extinction coefficient is fixed.

$$A_x/A_s = C_x/C_s$$

$A_x$ and $C_x$ mean the absorbance and the concentration of the tested solution, whilst $A_s$ and $C_s$ means the absorbance and the concentration of the standard solution of which the concentration is precisely known. Therefore, the concentration of unknown solution could be easily calculated mathematically.

However, the above method is limited in its precision. In practice, a standard curve of absorbance against concentrations is needed for measurement, which is plotted according to the absorbance of a series of standard solutions at known concentrations. Besides, the absorbance of standard solutions is measured under a given circumstance, such as a fixed wavelength of incident light and a fixed length of the light path, etc. With this standard curve, the concentration of an unknown solution can be shown according to its absorbance under the same situation.

Moreover, calibration of the spectrophotometer with a blank control, which contains the same solvent as all samples and no solute, is essential to for better quantification.

Spectrophotometry is an important technique used in many biochemical experiments that involve DNA, RNA, and protein detection, enzyme kinetics, and biochemical analyses. Besides quantification, spectrophotometry can also be used for a number of techniques such as determinations of optimal wavelength absorbance of samples, optimal pH for the absorbance of samples, and the p$Ka$ of various samples.

# Section 2    The Basic Construction of Spectrophotometers and It's Usage

## 2.1    The basic construction of spectrophotometers

Spectrophotometers employ the common basic components including a light source, wavelength selector ( monochromators) , slit, sample tube ( cuvette) and light–detecting system.

### 2.1.1    Light source

The light sourcein spectrophotometer must be capable of emitting a steady amount of light in a wide wavelength ranges. Most photometers employ a constant voltage–regulated tungsten lamp for spectral analyses in the range of 340–900 nm. More sophisticated spectrophotometers, which are capable for analysis in the UV range, are equipped with an additional constant voltage–stabilized hydrogen lamp that emits light in the range of 200–360 nm.

## 2.1.2   Monochromators

Photometry requires assay of absorbance at a given wavelength. Monochromators are employed to generate the desired wavelengths of light. Usually, it is not possible to produce light at a single wavelength. Therefore, when we speak of monochromatic light, we usually mean a source that has its maximum emission at this wavelength and has progressively less energy at longer and shorter wavelengths. The greater the spectral purity of monochromatic light, the higher the sensitivity and resolution of measurements.

The simplest wavelength selectors are one or more absorption filters that screen out the light above and below the specific wavelength. Some of the colorimeters are equipped with such filters. However, these filters generally have broad transmission ranges, and therefore the resolution of absorption spectra is limited. The modern monochromators containing a prism or a diffraction grating can overcome these deficiencies. Such monochromators generate relatively pure light at any wavelength over a wide range.

## 2.1.3   Slit

The intensity of light emitted through any filter or monochromators may be too intense or too weak for a light–sensing device to record. It is, therefore, necessary to adjust the intensity of the incident light ($I_0$) by placing a pair of baffles in the light path to form a slit. The desired intensity of light can be obtained by adjusting the width of the slit. Simple colorimeters often have a fixed slit, while sophisticated spectrophotometers usually have a variable slit mechanism.

## 2.1.4   Cuvette

Optical glass cuvettes are employed in visual range (400–700 nm) assays, and quartz cuvettes are applied in UV range (200–400 nm) assays. In order to keep the cuvette at good working conditions, the optical surface of the cuvette must be protected. Do not touch the surface with fingers or other rough things. After the assay, the cuvette must be rinsed with distilled water immediately, and do not leave any measuring solution in the cuvette after use, especially solutions of proteins and nucleic acids.

## 2.1.5   Light detecting system

Selenium photovoltaic cell, vacuum phototube, and photomultiplier tubes can be included as the photosensitive instruments through which the light energy can be converted into an electric signal. The current induced is proportional, within a certain range, to the intensity of light incident upon the instrument.

A colorimeter uses selenium photovoltaic cell as the light receptor. The cell is limited by low sensitivity and insensitivity to wavelengths shorter than 270 nm and longer than 700 nm. Accurate spectrophotometers usually use vacuum phototubes or photomultiplier tubes as the light detectors. Moreover, there are some instruments to amplify the photocurrent for increasing sensitivity.

## 2.2   Usage of the typical spectrophotometer

The spectrometer is widely used in laboratories. The specific instructions differ from other models, but the principles remain.

(1) Switch on the spectrophotometer at least 30 min before use, to allow it to warm up.

(2) Use the wavelength knob to set the desired wavelength for detection. Extreme wavelengths, in the

ultraviolet or infrared ranges, require special filters, light sources, and/or cuvettes.

(3) With the sample chamber closed, switch the mode to "T" (Transmittance).

(4) Calibrate the spectrophotometer with the reference cuvette (in the case, no reference cuvette is available, keep the sample chamber open wide), press the "0%" button to make a "0" on the % transmittance scale.

(5) Calibrate the spectrophotometer with blank control: put blank control in the light path, press the "100%" button to make a "100" on the % transmittance scale.

(6) Load about three-quarters of the total volume of sample solutions in each cuvette.

(7) Insert the cuvettes into the sample chamber, and switch to the "A" (Absorbance) mode.

(8) Sit sample cuvette in the light path one by one, record the absorbance data.

(9) Remove the sample cuvette, re-calibrate the spectrophotometer if necessary before checking the next sample.

(10) Clear and clean the sample cuvettes, switch off the spectrophotometer.

## 2.3 Notes

(1) Failure to allow the spectrophotometer enough time to warm up could result in imprecise results.

(2) Make sure cuvettes are free of any particles, smudges or fingerprints, as these can dramatically interfere with the machine's calculations.

*Hu Xiaojuan*

# Chapter 4

# Electrophoresis

Electrophoresis is the phenomenon that charged particles migrate directionally in the electric field. The particle moves towards the electrode with the opposite electrical property. Electrophoresis is also meant the technology used to isolate or purify the sample contains molecules with differences in size, charge, or conformation.

## Section 1    Basic Principle of Electrophoresis

## 1.1    Generation of charged particles

Substance canobtain charges by ionization or absorbing charges from other charged particles. The charged particle might be a metal ion, such as $Na^+$, or macromolecules such as protein, nucleic acid, etc.

A lot of amino acid residues in protein contain ionizable groups, such as $—NH_3^+$ and $—COO^-$. The ionization of these groups are sensitive to pH value changes of the solution. At isoelectric point (pI), protein contains and equal number of opposite charge. Therefore, it bears no net charge, termed zwitterion. So that protein is negatively charged at a pH higher than its pI, and is posilively charged at a louer pH.

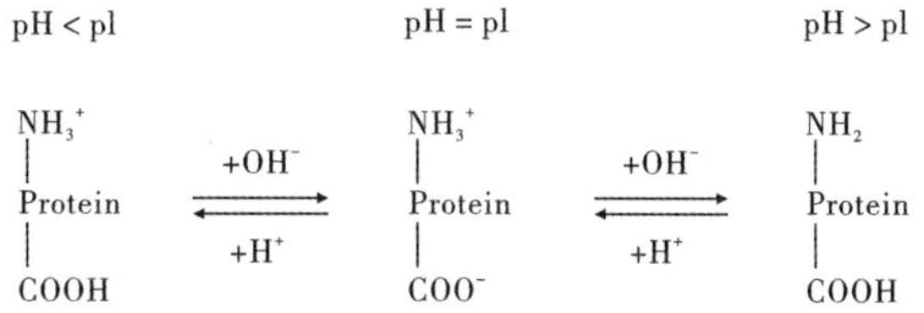

Figure 4.1    The electrolysis of protein in different pH

## 1.2  Electrical mobility

Electrical mobility is the ability of charged particles to move through a medium in response to an electric field. The electrical mobility of the particle is defined as the ratio of the drift velocity ($V$) to the magnitude of the electric field ($E$), $M = V/E$. The mobility of a particle is dependent on the property of the particle itself, including the quality of charges ($Q$), size ($r$), conformation ($\eta$), as the following formula shows.

$$M = \frac{Q}{6\pi r\eta}$$

The velocity of the particle is equal to mobility times electric strength. If different particles vary in mobility, when they move in the similar electric condition, velocity and distance of particles are different, so that they will be separated.

# Section 2   The Effect Factors of Electrophoresis

The charged particle migrates in the electric field, so that all the factors regulating the particle, electric field, and medium, will change the velocity of migration.

## 2.1  The quality of charges, size and conformation of particle

According to the formula of mobility, the velocity of a charged particle is positively correlated with the charges, negatively correlated with size. Generally, the spherical particle moves faster than fibrous one, because the medium viscosity of spherical particle is smaller than the fibrous one.

## 2.2  The electric buffer

The pH value, ionic composition, and concentration of electrophoresis buffer also influences the velocity of electrophoresis.

(1) pH value: for the ampholytes like protein and amino acids, the pH value of electric buffer defines their ionization. The larger the difference between target ampholytes' pI and pH value of the buffer, the more net charges those molecules carry and the faster the speed of migration. However, the protein will be denatured at in the extreme high or low pH, a suitable pH value of the buffer should be selected to make sure the difference of each sample's velocity is large enough to separate all the samples.

(2) Ionic composition: the components in the electric buffer must be very stable, and hard to electrolysis. General components include formate, acetate, phosphate, citrate, barbiturate, etc. It is necessary to choose a suitable buffer according to the sample. For example, the buffer of barbital–barbiturate is suitable for separating serum proteins.

(3) Concentration: the concentration of buffer generally refers to the ionic strength. If the ionic strength is high, the electricity carried by buffer will increase, while the electricity carried by the sample will decrease, which cause the particle migrating slowly. At the same time, the temperature of the system will increase, it does not benefit the separation effect. On the contrary, if the ionic strength is low, the velocity of

the sample will increase, and the heat production of the system is limited. However, the diffusion of the sample will be serious and it also not benefits the separation effect.

## 2.3 The electric field intensity

The electric field intensity refers to the electric potential gradient, which means the potential fall on each centimeter of the electric field. The higher the electric field intensity, the faster the particle migrates. However, the high voltage causes high electricity and more heat produced. High temperature leads to denaturation of proteins, decreased medium viscosity, increased thermal motion of particles, and the accelerated free diffusion. The high temperature also increases the ionic intensity and causes the siphonage. The temperature of the electric system should be controlled by regulating the voltage or current, and the cooling equipment also is needed to maintain a stable temperature when necessary.

## 2.4 The supporting medium

The ideal supporting medium should have inertness and not react to sample or buffer components. It also should be tough, hard to rupture and easy to preserve. So that the supporting medium is chosen according to the property of the sample. The common interaction of supporting medium and sample includes absorption and electroosmosis.

(1) Absorption: the surface of supporting medium absorbs the components in the sample, which makes the sample detained and expressed as the tailing. Since the absorption of the different component is different, which decrease the resolution effect of electrophoresis.

(2) Electroosmosis: electroosmosis is the motion of liquid induced by and applied potential across a porous material, membrane, or other supporting medium. It happens on the interface of supporting medium and liquid. For example, the filter paper is used as the supporting medium. The surface of the filter paper is negatively charged since hydroxyl groups are distributed. The liquid contacting the surface is positively charged and will migrate towards anode during electrophoresis. Electroosmosis certainly influences the velocity of the sample. If the directions of electroosmosis and migration of sample are similar, the migration of sample will speed up, and vice versa.

## Section 3   The Popular Electrophoresis Techniques

## 3.1 Cellulose acetate membrane electrophoresis (CAME)

Cellulose acetate is the acetate ester of cellulose. Cellulose acetate membrane, CAM, is the microporous membrane made of cellulose acetate fiber. CAME is a kind of zone electrophoresis utilizing CAM as the supporting medium, which can be widely used for separating several proteins. The resolution of CAME is low, and the loaded sample is too less to be used for preparing since the thickness of the membrane is just 10– 100 μm.

# 3.2   Agarose gel electrophoresis (AGE)

Agarose is a polysaccharide, generally extracted from certain red seaweed. It is a linear polymer made up of the repeating unit of agarobiose. The agarose in the gel forms a meshwork that contains pores, and the size of the pores depends on the concentration of agarose.

## 3.2.1   Characteristics of AGE

The advantages are as follows.

(1) The liquid in the agarose gel can be up to 99%, so AGE is similar to the free electrophoresis. However, the diffusion of the sample is less than free electrophoresis.

(2) The velocity of migration is high in AGE.

(3) The AGE also has high resolution and good repeatability.

(4) The agarose gel is transparent and does not absorb the ultraviolet so that it benefits quantitative determination by the ultraviolet detector.

(5) The zone in the gel can be dyed and the sample can be recycled for further research.

The disadvantage of agarose is the obvious electroosmosis effect because there are multiple sulfate radicals. AGE is generally used for separating and identifying nucleic acid. It's separation effect is depended on the size, conformation of nucleic acid, and the concentration of agarose.

## 3.2.2   The size and conformation of nucleic acid influences the velocity of migration

(1) Molecular conformation: DNA molecules have different conformations. For example, the plasmid DNA may be superhelical circular, linear, or with nicked circular structure when one strand is broken. If the molecular weight of plasmids is similar, the superhelical circular DNA migrates fastest, linear DNA takes the second place, and the nicked circular DNA migrates slowest.

(2) Size: generally, the smaller molecules migrate faster in the agarose gel. The mobility of the DNA fragment is negatively correlated with the logarithm of its molecular weight. The size of an unknown fragment could be measured by comparing the distance with standard fragments with known molecular weights. However, once the DNA is larger than 20 kb, it is hard to be separated by AGE.

# 3.3   Polyacrylamide gel electrophoresis (PAGE)

Polyacrylamide is a type of solid gel created by polymerization of acrylamide (Acr) linked by bisacrylamide (Bis) (Figure 4.2). A source of free radicals and a stabilizer, such as ammonium persulfate (AP) and (tetramethylethylenediamine, TEMED), is necessary to initiate polymerization. AP is the catalyst, and TEMED is the accelerator. PAGE is widely used for isolating and analysis of biomolecules such as protein, enzyme, and nucleic acids, etc. It is also used for the preparation of samples.

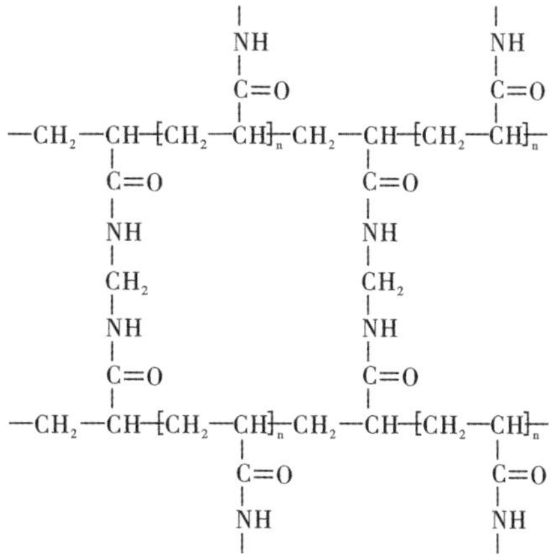

Figure 4.2   **The structure of polyacrylamide gel**

### 3.3.1   Characteristics of PAGE

(1) The polyacrylamide gel is transparent, flexible, and has good mechanical performance.

(2) The polyacrylamide gel is stable under different pH value or temperature. The chemical property of PAG is also stable and it does not react with the sample.

(3) PAGE has high sensitivity so that the dosage of the sample is small.

(4) The polyacrylamide has no ionized groups, it is not charged so that it is no absorption and electro-osmosis.

### 3.3.2   The pore size and property of PAG

(1) The property of polyacrylamide gel depends on the concentration of acrylamide and the degree of crosslinking. The total concentration means the amount of Acr & Bis in 100 mL gel solution and expressed as T%. The crosslinking degree is usually expressed as C%, which means the ration of Bis to the sum of Acr and Bis. The concentration of acrylamide can also be varied, generally in the range from 5% to 25%. The crosslinking degree can be varied for special purposes but is generally 2%–5%.

(2) The gel concentration is selected according to the size of the sample. Lower percentage gels are better for resolving higher molecular weight molecules, while much higher percentages of acrylamide are needed to resolve smaller proteins.

### 3.3.3   The basic principle of PAGE

In PAGE, the samples are isolated by condensation effect, charge effect, molecular sieving effect, etc. The charge effect means the different components in the sample isolated by their different charges since the mobility of the particle is proportional to the number of charges. The molecular sieving effect means the particles migrate in the porous gel and are isolated by different size. The bigger the particle is, the slower it migrates.

The PAGE can be prepared as a "continuous system" and "discontinuous system" (Figure 4.3).

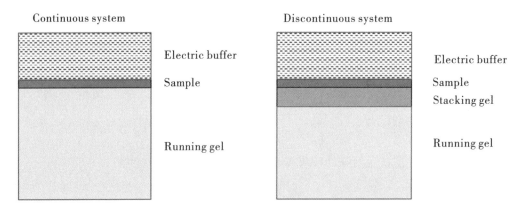

Figure 4.3    The different system of PAGE

The continuous system means the whole gel has a constant concentration, the sample moves in the constant environment. In a continuous system, particles are isolated only by charge effect and molecular sieving effect.

In a discontinuous system, the polyacrylamide gels are divided into two layers: the upper one is a macroporous gel with low concentration, called stacking gel, buffer for the formulation of this layer is Tris–HCl, pH 6.7; the lower one has higher concentration, called running gel, or separating gel, and the buffer for this gel is Tris–HCl, pH 8.9. Electrode buffer in the electrophoresis tank is Tris–glycine, pH 8.3 (Table 4.1). Obviously, the gel concentrations, compositions, pH and the electrophoresis buffer systems are different from stacking get to separating gel, thus forming a discontinuous system. In a discontinuous system, particles are separated via a condensation effect, charge effect, molecular sieving effect.

Table 4.1    The constitution of a discontinuous system

|  | Pore | pH | Ionic composition |
|---|---|---|---|
| Electric buffer |  | 8.3 | Glycine⁻ |
| Stacking gel | big | 6.7–6.8 | Cl⁻ |
| Running gel | small | 8.9 | Cl⁻ |

In the discontinuous system, samples will move across the stacking gel first. As soon as the power is turned on, glycine, proteins, chloride ions and bromophenol in HCl would be dissociated into anion, forming an ion flow and moving to the anode. when the glycine ions entered into the stacking gel, the pH value is low to 6.7 and close to its isoelectric point 5.97, so that the dissociation degree of glycine drops, the amount of charge reduced significantly and then the mobility became slower. The pIs of proteins in the sample are also closed to the pH value of the stacking gel, protein migrates slowly too. What's more, the pore size of stacking gel is too large to cause obstruction to protein molecular. Finally, in the stacking gel, the mobility of various ions is in the order of glycine⁻<protein⁻<Cl⁻.

The decline of dissociation degree after the glycine entering into the stacking gel makes the sudden absence of mobile ions flowing, resulting in reduced conductivity and electric current decline. However, the entire electric current of the other part of the electrophoresis system remains unchanged. On the basis that conductivity is inversely proportional to the potential gradient ($E = I/N$, $E$ stands for the potential gradient, $I$ stands for current intensity and $N$ stands for conductivity), there suddenly formed a high local potential gradient between Leading ion–CI and slow ion–glycine. Protein components in this local potential gradient region quickly migrate to the CI–ions region at different speed under the function of the high electric field. Through this process, the protein sample has been concentrated for several hundred folds and the protein components are arranged in a certain order to form a layer.

When the ion flow continued to move forward and entered into the running gel prepared by the pH 8.9 buffer, the protein molecule encountered resistance. Then the mobility became slow. At the same time, under the conditions of pH 8.9, glycine would fully dissociate. Its electricity would increase, eliminating the phenomenon of ion missing. Each section of the gel recovered with a constant electric strength. Proteins begin to migrate at different rates, because of the sieving properties of the gel. Smaller protein SDS complexes migrate more quickly than larger protein SDS complexes. Within a certain range determined by the porosity of the gel, the migration rate of a protein in the running gel is inversely proportional to the logarithm of its MW.

Seen from the principle above, the main advantages of discontinuous polyacrylamide gel electrophoresis is that when the protein samples go through the stacking gel, they can form a tightly compressed layer and flow into the separating gel. With the protein components separated previously and compressed into a layer, it can reduce the interference caused by the zone overlapping, thus improving the distinguish ability of electrophoresis. According to the above analysis, protein samples would be well separated.

## 3.3.4　Derived electrophoresis based on PAGE

(1) Disc or slab electrophoresis

This two electrophoreses share a similar principle and employ different apparatuses. The disc electrophoresis is named after that the isolated and dyed samples are shown as a disc since the gel is polymerized in a thin glass tube. The slab electrophoresis utilizes two glass slabs to prepare the gel, and after electrophoresis and dying, the samples are shown as bands.

By contrast, slab electrophoresis has more advantages. Over 10 samples could be detected in a unique gel and it is convenient to analyze in the same condition. And the sample can be transformed into the membrane for blotting or detected by autoradiography. The gel is thin, generally, like 0.5 mm, 1.0 mm, or 1.5 mm, the loading quantity of the sample is small, and the resolution is high. The superficial area of the gel is large and convenient to cooling so that the bands are clear. It is easy to prepare the gel and peeling it. The gel can be preserved and scanned after dying and drying.

(2) SDS-PAGE

Once sodium dodecyl sulfate, simply expressed as SDS, is added in the electrophoresis system, the mobility of particles depends mainly on the molecular weight, and have no concern with net charges and conformation, this method is called SDS-PAGE.

SDS is an anionic detergent, which can bind with protein to form SDS-protein complex. Since there are lots of negative charges in SDS, the negative charges carried by SDS exceed those carried by protein, the difference of charges in proteins are eliminated or covered up.

In addition, in SDS-PAGE, SDS and β-mercaptoethanol are routinely added in the loading buffer. β-mercaptoethanol reduces the disulfide bond in the protein and dissociate the polymer to monomers. SDS binding breaks the hydrogen bonds and hydrophobic bonds in the protein, which causes the change of conformation of the protein. The SDS-protein complex is always clublike with the identical minor axis about 1.8 nm, and the long axis is proportional with the molecular weight of protein.

In conclusion, the mobility of the SDS-protein complex is not relative to charges and conformation of the protein, and only relative with the molecular weight of protein.

(3) Isoelectric focusing (IEF)

Isoelectric focusing (IEF), also known as electrofocusing, is a technique for separating different molecules based on the differences in their isoelectric points. IEF involves the acrylamide gel matrix co-polymerized with the pH gradient, which results in completely stable gradients. The proteins migrate in the gel and stop at the region with pH value equals to its pI. As a result, the proteins become focused into sharp stationary bands with each protein positioned at a point in the pH gradient corresponding to its pI. The technique is capable of isolation of different protein with similar molecular weight and different pI points. The resolution is extremely high to the 0. 01 pH unit.

(4) Two-dimensional electrophoresis (2-DE)

2-DE is a combination of with IEF and SDS-PAGE. In the first dimension, molecules are separated by IEF linearly according to their isoelectric point. In the second dimension, the molecules are separated at 90 degrees from the first electropherogram according to molecular mass (Figure 4. 4). Since it is unlikely that two molecules will be similar in two distinct properties, molecules are more effectively separated in 2-D electrophoresis than in 1-D electrophoresis. It is the core technology in proteomics research.

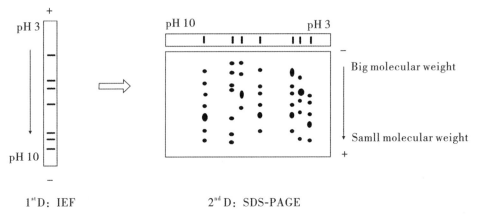

Figure 4. 4   The schematic diagram of 2-DE

# Section 4   Dying Method

## 4.1   Staining of protein

### 4.1.1   Amino black 10B staining

Amino black 10B possesses an acidic dye-containing sulfonic group, which can react with protein. It stains the proteins a blue-black color. The disadvantages of amino black 10B are that the degree of staining is varied in different proteins, and the staining is not stable. It is not fit for scanning and analysis of gray.

### 4.1.2   Coomassie brilliant blue (CBB) staining

That is one of the most popular stains. CBB binds with protein by Van der Waals' force. There are two CBB, R250 and G250. CBB R250 staining has higher sensibility. CBB G250 contains more 2 methyl groups than R250 so that the sensitivity of G250 is lower than R250. However, G250 is not soluble in the trichloroacetic acid and form a colloid, which can selectively stain protein without background coloration, so that the G250 staining is stable and repeatable, and is applicable to quantitative analysis.

### 4.1.3   Silver staining

The silver ion binds with multiple groups, such as sulfhydryl group, carbonyl group. Under the alkaline condition, the silver ion is reduced to metallic silver, which subsides on the surface of the protein and develops color. This staining method has high sensitivity, which can stain the point of protein less than 1 ng in the gel so that is widely used in the 2−D gel and gel with extremely low protein.

# 4.2   Staining of lipoprotein

The popular dyes for lipid or lipoprotein are Oil Red O and Sudan Black B. Oil Red O is a lysochrome (fat−soluble dye) diazo dye used for staining of neutral triglycerides, lipids, and some lipoproteins. Oil Red O provides deep red color and the stains are easy to be seen. Sudan Black Bis a nonfluorescent, relatively thermostable lysochrome diazo dye. It stains lipids blue−black. Sudan Black B can be used to stain some other materials, as it is not so specific to lipids.

# 4.3   Staining of nucleic acid

## 4.3.1   Ethidium bromide(EB)

EB is the most commonly used dye for nucleic acid detection ingels. EB is a DNA intercalator by inserting itself between the base pairs in the double helix. Once exposed to ultraviolet light, it will fluoresce with an orange color. After electrophoresis, when the agarose gel is illuminated using UV light, the nucleic acid bands become visible. The property of EB includes high sensitivity, simple operation, without breakage of DNA, etc.

However, EB may be a strong mutagen. UV is also harmful to eyes and skin. Individual protection is necessary during the operation and observing, such as gloves, eyeglass mask, etc.

## 4.3.2   Methyl green

Methyl green is apositively charged stain with high affinity for double−stranded DNA and shows green under UV light. Unlike EB, its interaction with DNA has been shown to be non−intercalating, not inserting itself into the DNA, but instead electrostatic interaction with the DNA major groove. It is usually used to detect the natural DNA.

## 4.3.3   Diphenylamine

In acidic condition, the deoxyribose of DNA is converted to $\omega$−hydroxy−$\gamma$−keto−valeraldehyde, which binds with diphenylamine and shows blue. The depth of blue is proportional to the amount of DNA. This method can be used to distinguish DNA and RNA.

## 4.3.4   Pyronine Y

Pyronine Y has high affinity to RNA. Once binding with RNA, it shows red. This method has good staining effect and high sensitivity.

*Yuan Ping*

# Chapter 5

## Chromatography

Chromatography is a laboratory technique for the separation of the multi-component mixture, based on the physical, chemical, and biological differences. As an important method of analysis and separation, chromatography is widely used in scientific research and industrial. It plays an important role in the fields of petrochemical, pharmaceutical, biological science, environmental science, and agricultural science.

## Section 1    Basic Principle and Classification

### 1.1    The basic concept of chromatography

(1) Stationary phase: stationary phase is a matrix of the chromatography system. It can be a solid substance (such as an adsorbent, a gel, an ion exchange agent), or a liquid substance (such as the solution fixed to silica gel or cellulose). The matrix is capable of reversible adsorption, dissolution, and exchange of compounds which are to be separated.

(2) Mobile phase: a mobile phase is a liquid or gas which pushes a substance to be separated on a stationary phase and to move in one direction during the chromatography. In column chromatography, it is generally called eluent, and thin layer chromatography is called the spreading agent.

(3) Distribution-coefficient: the concentrations ratio of one compound in a mixture of two immiscible phases. The distribution coefficient is mainly related to the properties of the material being separated, the nature of the stationary phase and mobile phase, the column temperature and so on. Under certain conditions, it is constant and denoted as $K$ (Equation 5.1). The constant $K$ has a different meaning for the different type of chromatography, which can be regarded as adsorption equilibrium constant, distribution constant, or ion exchange constant, etc.

$$K = \frac{C_s}{C_m} \qquad \text{(Equation 5.1)}$$

$C_s$ is the solute concentration in the stationary phase; $C_m$ is the solute concentration in the mobile phase.

(4) Retardation factor ($R_f$): in planar chromatography, in particular, the retardation factor ($R_f$) is defined as the ratio of the migration distance of one substance to the migration distance of the solvent front (Equation 5.2).

$$R_f = \frac{\text{the migration distance of substance}}{\text{the migration distance of solvent front}} \qquad \text{(Equation 5.2)}$$

Generally, the $R_f$ value will always be in the range 0–1. It is mainly determined by the distribution–coefficient of a certain substance. The component with a larger distribution coefficient moves slower, and its $R_f$ value is smaller. On the contrary, the components with the smaller distribution coefficient have faster–moving speed and larger $R_f$ value (Figure 5.1). Since the distribution coefficient of each substance in a certain solvent is constant, its moving speed is also constant. Therefore, the different components can be identified based on the different $R_f$ value.

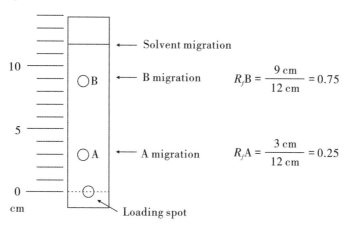

Figure 5.1   The component with the larger $R_f$ value moves faster

(5) Resolution ($R_s$): the resolution represents the degree of separation between two adjacent peaks (Figure 5.2). It can be calculated by Equation 5.3.

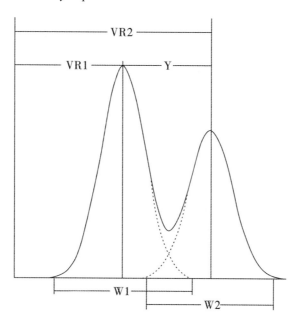

VR1: The total volume of eluent from component 1 to the corresponding elution peak. VR2: The total volume of eluent from component 2 to the corresponding elution peak. W1: Elution peak width of component 1. W2: Elution peak width of component 2. Y: Difference between the total volume of eluent at peak elution of component 1 and component 2.

Figure 5.2   The resolution of chromatography

$$R_s = \frac{VR_2 - VR_1}{\dfrac{W_1 + W_2}{2}} = \frac{2Y}{W_1 + W_2}$$

( Equation 5. 3)

The larger the $R_s$ value is, the better the separation of the two components is. When $R_s = 1$, the two components have good separation, about 2% of each other, that is, the purity of each component is about 98%. When $R_s = 1.5$, the two components are completely separated, and the purity of each component can reach 99.8%.

## 1.2   The basic principle of chromatography

The chromatographic system consists of a stationary phase and a mobile phase ( liquid or gas). The mobile phase moves relative to the stationary phase to cause the sample of the mixture to be separated to move forward through the stationary phase. Due to differences in physical, chemical and biological properties ( such as solubility, adsorption capacity, stereochemistry, and molecular size, polarity, affinity and specific biological response, etc. ) of the components of mixture samples, their ability to interact with each other is different. The component with a weak interaction with the stationary phase is less resistant to the mobile phase, and the moving speed is faster. On the contrary, the moving speed is slow. After a certain period of time, the different components in the mixture can be separated on the stationary phase.

If the distribution coefficient $K$ of a substance between the stationary phase and the mobile phase is 1, the concentration of this substance in the stationary phase is the same as in the mobile phase; if it is 0.5, the concentration in the mobile phase is two times in the stationary phase. The most fundamental reason why a variety of substances can be separated is that they have different distribution coefficients between the two phases.

The rate of migration of a substance in a column is determined by the distribution coefficient. The larger the distribution coefficient of the substance combined to the stationary phase, the slower it migrates. Conversely, the binding force is smaller, the faster it migrates. Therefore, based on the different distribution coefficient of each component in the mixture, they can be separated by choosing suitable chromatography system.

## 1.3   The classification of chromatography technology

According to the morphology of the stationary phase and mobile phase, as well as the separation principle, the chromatography can be divided into many types.

### 1.3.1   Classification according to the two-phase morphology

( 1) According to the properties of the stationary phase matrix: chromatography can be divided into paper chromatography, thin layer chromatography, and column chromatography. Paper chromatography refers to chromatography using filter paper as a matrix. Thin layer chromatography is the smooth surface of glass or plastic coated with a thin layer of the matrix for chromatography. Column chromatography refers to the filling of a matrix into a tubular container for chromatography in a column. Paper chromatography and thin layer chromatography are mainly suitable for the rapid detection, analysis, and preparative separation of small molecules on a small scale, which are usually disposable. Column chromatography is suitable for sample analysis, separation, and preparation. Commonly used gel chromatography, ion-exchange chromatography, affinity chromatography, and high-performance liquid chromatography ( HPLC) are usually column chromatography.

(2) According to the morphology of the mobile phase: chromatography can be divided into liquid chromatography (LC) and gas chromatography (GC). Gas chromatography requires the sample gasification, limiting its application, is mainly used for amino acids, nucleic acids, carbohydrates, fatty acids, and other small molecules analysis and identification. Liquid chromatography is suitable for the analysis and separation of biological samples. It is the most commonly used chromatography technique in the life science field.

## 1.3.2   Classification according to the separation principle

The classification chromatography technology is shown in Table 5.1.

Table 5.1   The classification chromatography technology

| Chromatographic type | Separation principle |
| --- | --- |
| a. Adsorption chromatography | The adsorption capacity of each component on the adsorbent surface is different. The stationary phase is usually solid adsorbent |
| b. Partition chromatography | The distribution coefficient of each component in the mobile phase and the stationary phase is different |
| c. Ion–Exchange chromatography | The stationary phase is an ion exchanger. The components own different affinities with the ion exchanger |
| d. Gel–Filtration chromatography | The stationary phase is a porous gel. The blocking effect on the gel is different due to the different molecular size of each component |
| e. Affinity chromatography | The stationary phase can only interact with one component to separate the other non–affinity components |

# Section 2   Adsorption Chromatography

## 2.1   The principle of adsorption chromatography

Adsorption chromatography is used for the separation according to the different adsorption properties between the adsorbent surface and the mixture materials. Since the difference of the physical properties of each component in the mixture (such as solubility, adsorption capacity, molecular shape, size, and polarity), different components of the mixtures have different adsorption affinity. Thus, when the mobile phase accompanied by the mixture passes through the stationary phase, the migration speed of each component is different, and different components will eventually be separated. The efficiency of adsorption chromatography depends on both the ability of the unseparated substances to be adsorbed by the adsorbent (stationary phase) and the differences of their solubility in the solvent (mobile phase) used during the separation.

There are many kinds of adsorbents in the stationary phase of adsorption chromatography. Inorganic adsorbents are alumina, activated carbon, silica gel and so on; organic sorbents such as cellulose, starch, sucrose and so on. Alumina and silica gel are the most commonly used. The above material has the properties of the adsorption of certain substances, and the absorption capacity of different substances is different. The adsorption affinity depends on the adsorbent itself and is also related to the nature of the adsorption material. The molecules (ions or atoms) on the solid surface of the stationary phase are not equal to the attraction of the solid internal molecules so that different substances can be absorbed on the solid surface. The adsorption process is reversible, and the substance absorbed can be desorbed under certain conditions.

## 2.2  The classification of adsorption chromatography

According to the different operating methods, the adsorption chromatography can be divided into two types: column chromatography and thin layer chromatography (TLC).

### 2.2.1  Column chromatography

Column chromatography is to use the glass column loading adsorbent (stationary phase) for the separation of the mixture. The column used for column chromatography is commonly a glass tube with the appropriate size. The bottom of the glass tube sets fine nylon mesh, glass wool or a suitable pore filter in order to pack the stationary phase in the column.

An impure sample is loaded onto a column of an adsorbent, such as silica gel or alumina. An organic solvent or a mixture of solvents (the eluent) flows down through the column. Due to different distribution coefficient, components of the sample separate from each other by partitioning between the stationary phase (silica or alumina) and the mobile eluent. Molecules with different polarity partition to different extents, and therefore their migration rates moving through the column are different. The eluent is usually collected in fractions, which are typically analyzed by thin layer chromatography to see if the separation of the components was successful.

In general, non−polar and less polar organic substances, such as β−carotene, glycerides, phospholipids, and cholesterol, etc., are much suitable to be separated by this method.

### 2.2.2  Thin layer chromatography (TLC)

Thin layer chromatography (TLC) is a chromatography technique used to separate non−volatile mixtures. Thin layer chromatography is performed on a sheet of glass, plastic, or alumina foil, which is coated with a thin layer of adsorbent material. Silica gel, alumina oxide, or cellulose usually used as the stationary phase. To perform TLC, a solution of a compound or mixture of compounds is applied to a TLC plate. Then TLC plate is carefully placed into an airtight chromatography chamber containing the solvent. The solvent level should not be above the sample spots on the plate. Allowing the solvent to climb up the TLC plate, and remove the plate when the solvent nearly reaches the top. Similarly, with column chromatography, each component can be isolated while the solvent climbing up the process due to their different distribution coefficient and displaying different retardation factor ($R_f$).

The TLC plate is prepared by mixing the adsorbent, such as silica gel, with a few inert binders like cellulose sodium or calcium sulfate and water. This well mixed thick slurry is spread on an unreactive carrier sheet, usually glass, thick aluminum foil, or plastic. The resultant plate is dried and activated by heating in an oven for thirty minutes at 110 ℃. The thickness of the absorbent layer is typically around 0.10 − 0.25 mm for analytical purposes and around 0.5 − 2.0 mm for preparative TLC. TLC plates are also commercially available.

Thin layer chromatography has many advantages, such as easy operation, fast stratification, high sensitivity, good separation effect, and the convenient choice between different stationary phases. In addition, it is convenient to display sample spot, and also possible to use a corrosive chromogenic agent, or even at a high temperature. Fluorescent dyes can also be added to the adsorbent material to aid in the identification of samples in different spots(Figure 5.3).

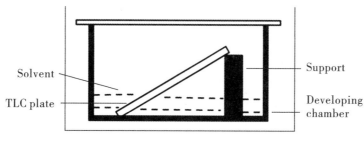

Figure 5.3　TLC sketch map

# 2.3　Adsorbent and elution

The selection of adsorbents and eluents is the key to adsorption chromatography. However, there is no consistent rule to select the adsorbents, which need to be screened. In general, the maximum specific surface areas and sufficient adsorption capacity are necessary parameters for the selected adsorbent. The adsorption capacity for each component should be different, and no chemical reaction with the solvent as well as each component. In addition, the adsorbent particles are also required to be uniform with high mechanical strength.

Alumina and silica gel are widely used as the polar adsorbents for chromatography. The water content of the silica gel and alumina has a great influence on their adsorption capacity, which can be divided into five levels (Table 5.2). At a certain temperature, the silica gel and alumina can be heated and dried to enhance the adsorption capacity, known as activation. Conversely, the addition of a small amount of water can decrease the activity, known as the deactivation. But the temperature should be controlled during the activation process. The too high temperature would cause the destruction of the internal structure of the adsorbent and lead irreversibly decline in adsorption capacity.

The silica gel is usually dried and activated by heating in an oven for thirty minutes around 110 ℃ (no more than 200 ℃).

To prepare activated alumina (grade I – II), the alumina is placed in an unreactive carrier sheet, and make sure the alumina thickness bellowing 3 cm. The alumina is typically heated in an oven for 6 h around 400 ℃, then placed in a sealed dryer and cooled to room temperature gradually.

Table 5.2　The correlation between water content and adsorption activity of silica gel and alumina

| Activity degree | The water content of silica gel (%) | The water content of alumina (%) |
|:---:|:---:|:---:|
| I | 0 | 0 |
| II | 5 | 3 |
| III | 15 | 6 |
| IV | 25 | 10 |
| V | 38 | 15 |

For hydrophilic adsorbent (such as silica gel, alumina): in order to separate the polar components, adsorbent with weak adsorption capacity (less active) and more polar eluent should be used; on contrary, adsorbents with strong adsorption capacity and less polar eluents should be selected to separate non – polar substances.

Polarity order of commonly used eluents: petroleum ether < cyclohexane < carbon tetrachloride < benzene < toluene < ether < chloroform < ethyl acetate < n−butanol < acetone < ethanol < methanol < water.

Adsorption capacity order of commonly used adsorbent: sucrose < starch < talcum powder <sodium carbonate < potassium carbonate < calcium carbonate < magnesium carbonate < magnesium oxide < silica gel < magnesium silicate < alumina < activated carbon < Fuller's earth.

# Section 3   Partition Chromatography

## 3.1   The principle of partition chromatography

Partition chromatography is a method of separating substances based on their solubility and partition coefficient between two immiscible ( or partially miscible) solvents. Most of the currently used partition chromatography techniques employ a porous material to sorb a polar solvent, which is referred to as the stationaryphase and is always immobilized on the porous support ( carrier) during the chromatography. Another non−polar and fixed immiscible solvent flows through the stationary phase, this mobile solvent known as the mobile phase.

Silica gel can absorb the equivalent of 70% of its own weight of water, then its adsorption properties disappear, and silica gel is a carrier for water and the water on the carrier is a stationary phase ( The carrier acts only as a stationary phase in chromatography. They are inert substances with low adsorption and weak reactivity, such as starch, cellulose powder, and filter paper) . In addition to water, the stationary phase may also have a strong polar solution such as dilute sulfuric acid, methanol, secondary amides, etc. However, the carrier must be capable of binding with a polar solvent in a non−flowing state.

The selection of mobile phase: in general, the solvent of each component may be selected as the mobile phase, and then according to the separation of the mobile phase to change the composition, you can add some other solvents in the mobile phase to form a mixed solvent to change the separation of the components and the elution rate. Commonly used mobile phase petroleum ether, alcohols, ketones, esters, alkyl halides, and benzene, etc. , or their mixture elute.

Partition chromatography method overcomes the difficulties encountered in adsorption chromatography. For example, polar substances such as fatty acids and polyhydric alcohols are strongly adsorbed by conventional adsorbents, which are difficult to elute and cannot be separated by adsorption chromatography. The partition chromatography method is very easy to separate polar substances such as the fatty acids. Therefore, this method is rapidly and widely used in the separation of polar organic compounds. At present, the distribution chromatography is mainly used for the separation of polar and hydrophilic substances such as organic acids, amino acids, carbohydrates, peptides, nucleosides, and nucleotides.

## 3.2   Paper chromatography

According to the different supporting materials used for stationary pharse, the distribution chromatography can be divided into paper chromatography ( paper distribution chromatography) , column chromatography and thin layer chromatography. Paper chromatography is the most widely used distribution chromatography, and it has become an important separation and analysis tool in biochemical research. It is very useful for the isolation and identification of small molecules such as amino acids, peptides, nucleosides, nucleotides, sugars, and vitamins. The following focuses on the paper chromatography, which is actually very similar to thin layer chromatography. Its operation methods (including the sample, exhibition, and color, etc.) is basically the same as that of TLC.

### 3.2.1  The principle of paper chromatography

Paper distribution chromatography is based on filter paper as an inert support. Filter paper fibers and water have a strong affinity; can absorb about 22% of the water. However, filter paper fibers have a low affinity with organic solvents, so filter paper can be regarded as aqueous inert support and water as a stationary phase. Some organic solvents such as alcohols, phenols are common mobile phases.

The substance to be separated is added to one end of the paper and the mobile phase is passed through the sample point. The solute on the sample must be distributed between the aqueous phase (stationary phase) and the organic phase (mobile phase). As the organic phase continues to flow, the solute moves along the flow of the organic phase and is continuously distributed. If the solute is more soluble in the stationary phase, the solute moves slower with the mobile phase, and vice versa. The solute is separated due to the different moving speed of different components.

### 3.2.2  The type of paper chromatography

By method of operation, paper chromatography can be divided into vertical type and horizontal type. Vertical paper chromatography is to suspend the filter paper, making the mobile phase spread upward or downward. Horizontal paper chromatography is the circular filter paper placed in the horizontal position, making the solvent spread from the center to the surrounding.

Vertical paper chromatography is widely used, and according to the amount of material separated, the filter paper is cut into strips, the sample is point at one end away from the edge of 2−4 cm, to be dry, making the spot−like end edge of the solution in contact with the solution, expanding in a covered glass jar (Figure 5.4).

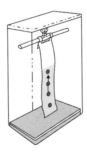

Figure 5.4  Vertical paper chromatography

The above method using only one solvent system for one expansion is called unidirectional chromatography. If the sample contains more components, and their $R_f$ is similar, unidirectional chromatography separation is not effective. Thus, two−dimensional chromatography can be used that is spotting at the rectangular or square corner of the filter paper, rolled into a cylindrical. Firstly, be developed with one solvent system, dried; secondly, be rotated 90° and placed in another solvent system, and then be developed in the other direction. So that the components are separated more clearly (Figure 5.5).

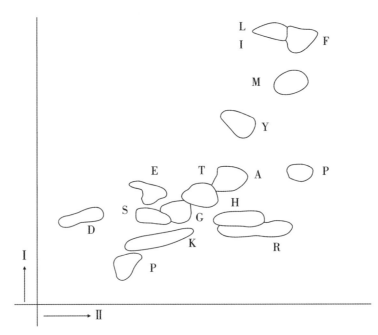

Figure 5.5    The separation of amino acids by two–dimensional paper chromatography

# Section 4    Ion-exchange Chromatography

## 4.1    The principle of ion–exchange chromatography

Ion–exchange chromatography ( IEC) is a separation chromatography using ion exchanger as the stationary phase. The interaction between ions in the mobile phase and the ion exchanger is different in the reversible exchanging process, and lead to effective separation.

The stationary phase of the IEC ion exchanger is usually a water–insoluble polymer, such as resins, cellulose, dextran, agarose, etc. They are insoluble in water, acidic or basic solution, and stable to heating, organic solvents, oxidation or reduction as well as other chemical reagents. The active chemical groups ( commonly as an acidic or basic group) , bonded to the ion exchanger, can dissociate and generate ions that exchanging with the sample in solution under certain conditions.

Ions in the sample solution can be reversibly adsorbed and dissociated with the ion exchanger in different pH and ionic strength solution. For example, the different isoelectric point ( pI) , charge number, and molecular size for each protein can lead to a different number of ionic bond with the ion exchanger. Therefore, the interaction between protein and the ion exchanger is much different. More ionic bonds lead to stronger interaction, and shorter moving distance, and longer retention time on the top of the column during the chromatography. On the contrary, less ionic bonds favor the dissociation of proteins with the stationary phase and make the protein more easily flowing down follow with the solvent, thus the separation of different samples can be achieved.

# 4.2    The type of ion–exchanger

The commonly used ion–exchanger is polymer compounds, including ion-exchange resins, ion-exchange cellulose, ion-exchange dextran, ion-exchange agarose and so on. Taking ion-exchange resin, for example, its classification is described as followings.

According to the different properties of the ion dissociated from the exchanging groups on the stationary phase, ion-exchange resin can be divided into cationic exchange resin and anion exchange resin. In addition, either cationic or anionic exchange resin can also be divided into strong or weak type.

Cations (such as $H^+$) can be dissociated via introducing the acidic groups (such as $—SO_3H$) to the polymer. These cations can be exchanged with the cations in the solution. This type of polymer is called cationic exchange resin. When introducing the basic groups (such as quaternary ammonium) to the polymer, anions (such as $OH^-$) can be generated and exchanged with anions (such as $Cl^-$) in the solution. Such resins are referred to an anionic exchange resin. The ion-exchange resin can be shown as follows.

$[Resin]—SO_3^-……H^+$                  $[Resin]—CO_2^-……H^+$

Strong Acidic Cationic Exchange          Weak Acidic Cationic Exchange Resin

$[Resin]—N^+R_3……OH^-$               $[Resin]—NH_3^+……OH^-$

Strong Basic Anionic Exchange Resin    Weak Basic Anionic Exchange Resin

The ion–exchange reaction is shown as follows:

$$R–SO_3^-\ H^+ + M^+\ X^- \longrightarrow R–SO_3^-\ M^+ + H^+\ X^-$$

$$R–N^+R'_3OH^- + H^+\ X^- \longrightarrow R–N^+R'_3X^- + H_2O$$

The ion-exchange reaction is an equilibrium process. While continuously adding the exchange solution in the column, however, the equilibrium can proceed continuously in the positive direction until all the ions in the ion–exchangers being completely eluted (Figure 5.6). Similarly, when a certain amount of solution flows through the exchange column, the ion concentration of the solution will be gradually decreased due to the continuously ion exchanging process. Thus, the certain ion can also be adsorbed on the resin.

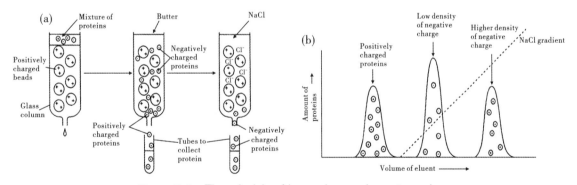

Figure 5.6    The principle of ion-exchange chromatography

# 4.3    The application of ion–exchange chromatography

Ion–exchange chromatography is widely used in the separation and purification of small molecules (such as amino acids, nucleotides, antibiotics, organic acids, inorganic ions). The preparation of "deionized water" is a typical example. Many types of cations (such as $K^+$, $Na^+$, $Ca^{2+}$, and $Mg^{2+}$), as well as anions (including $Cl^-$, $Br^-$, $SO_2^{2-}$, $HCO_3^-$), are contained in natural water. In order to remove these ions, natural water should pass through the $H^+$-type cationic exchange column and then pass through $OH^-$ anionic exchange column, and then the obtained effluent is "deionized water".

Practically, the appropriate ion-exchangers should be chosen according to the physical and chemical properties of the compound to be separated. Polystyrene ion-exchangers are commonly used for the separation of small molecules such as inorganic ions, amino acids, nucleotides and so on. Polystyrene ion-exchanger owns advantages such as good mechanical strength and fast flow rate. However, its weak affinity with water and highly hydrophobic properties easily lead to protein denaturation. Therefore, the stationary phase with a stronger affinity with water, such as cellulose, spherical cellulose, dextran, and sepharose, are usually explored in protein separation.

# 4.4    The preparation of ion-exchange resin

## 4.4.1    New resin pretreatment

The new resin is usually dry resin so that it needs to be fully hydrated and swollen in water. As impurities being contained, the resin needs to be washed with water, acidic as well as the basic solution. The general procedure is as follows: the new factory-made resin is firstly soaked in water for 2 h. After removing small resin particles floating in the solution, the bubbles should also be removed under vacuum pumping. The resin is then washed with plenty of deionized water. After adding the hydrochloric acid solution (2 mol/L) and stirring for 4 h, the acidic solution is removed and thoroughly washed with water to neutral. A similar procedure is repeated but using NaOH solution (2 mol/L), and washed to neutral. The resin needs to be balanced before use buffer with the required pH and ionic strength.

## 4.4.2    The general role of column loading

The height and diameter ratio of the column is better around (10-20) : 1. The gravitational sedimentation method is commonly used. The key issue is that the resin must be distributed uniformly in the column to prevent out-of-phase and generation of bubbles. In addition, the surface of the column must be smooth.

## 4.4.3    Elution and collection

The eluent selection is different for the separation of different samples. The principle is using a more active ion to exchange the adsorbed ions. Since the adsorbed components are always not a single substance that required. Controlling the flow rate and stepwise collection are also key issues to obtain a single substance, besides choosing the proper eluent.

## 4.4.4    Regeneration

The method of restoring the used resin to its original state is called regeneration. However, it is no need to wash the resin with acidic as well as a basic solution for every round of regeneration, but only the transformation process is needed. The transformation process refers to the operation of loading the certain ion to the resin. For example, the $Na^+$-type ion-exchange resin can be generated with stirring the resin in four-fold amount (volume) of NaOH solution for over 2 h, and the $H^+$-type ion-exchange resin can be generated and treated with HCl solution. Similarly, the transformation of anionic exchange resin is following the similar procedure, which is $Cl^-$-type and $OH^-$-type ion-exchange resins are treated with HCl and NaOH, respectively.

# Section 5   Gel Chromatography

Gel chromatography refers to a technique in which the mixture is passed through the column packed with gel as the stationary phase, and separated by the molecular size. Due to the entire process of the gel chromatography is similar to filtration, it is also known as gel filtration or molecular sieve filtration. Gel chromatography is developed in the early 1960s, and it is a rapidly developed and simple separation technique. It owns a series of advantages, including simple device, highly effective, easily handled, and so on. It is mainly used for the separation of protein, nucleic acid, and other macromolecules or polymer material.

## 5.1   The principle of gel chromatography

The gel is dry and particulate material. It can be called gels, only after being soaked in a suitable solvent to absorb a large amount of liquid and swell well. Because the gel is a kind of cross−linked polymer with a three−dimensional structure, it also has significant molecular sieve effect.

After putting appropriate gel particles into the glass tube to make a chromatography column and adding the sample mixture to the column, the chromatography can be preceded via eluting the column with a large amount of distilled water or other dilute solution. Due to the different molecular size and shape of the substances in the mixture, during the washing process, substances with higher relative molecular weight cannot flow into the interior of the gel meshes but only flow with the solvent along the gap between the gel particles. This component moves fast and is the first to be eluted out of the column, due to its weak resistance that it encountered. A substance with lower relative molecular weight can get into the interior of the gel mesh, and cause its slow−moving rate, and be eluted out of the column in late stage. After collecting the effluent in a different stage, substances with different molecular weights can be separated (Figure 5.7).

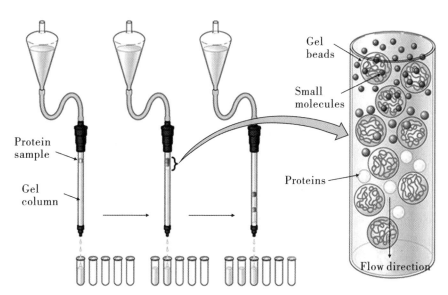

Figure 5.7   The principle of gel filtration chromatography

Gel filtration chromatography can be divided into two types. Firstly, in the separation of substances with a high difference in the relative molecular weight, the material with high relative molecular weight cannot enter gel at all, and would be eluted out of the column. The materials with low relative molecular weight are fully penetrated by the gel. This is called group separation, commonly used for preparative separation, such as desalting. Another situation is the separation of the substances with similar relatively molecular weight, and the resolution is higher than the group separation, is called fractionation, commonly used in the analysis.

# 5.2   Experimental technique of gel chromatography

## 5.2.1   Gel selection

The gels used in gel chromatography include Sephadex, polyacrylamide gel, and agarose gel. Now, the most commonly used gels are cross–linked dextran gels. It has eight –types (Table 5.3, known as "G–type"), which indicates the cross–linked grade. The more highly cross–linked, the denser the reticular structure, the smaller the "mesh", and the less the amount of water absorbed, which is suitable for the separation of substances with lower relative molecular weight. The model of the commercially available gel is generally expressed using the number which is 10 times of the absorbed water amount. G–25 type of gel means 1 g of gel can absorb 2.5 g of water. The gel with a water absorption over 7.5 g per gram is called a soft gel, and the gel with water absorption less than 7.5 g per gram is called a hard gel.

The amount of dry gel required is estimated based on the desired gel volume. Generally, the volume of dextran gel after the absorption of water is about double the amount of water volume. For example, the water absorption of Sephadex G–200 is 20 mL per gram, and its final volume is about 40 mL.

Table 5.3   The different type and characteristic of Sephadex gel

| Model | Molecular weight | Water absorption (mL/g) | Soaking time(h) 20–25 ℃ | Soaking time(h) 100 ℃ | Sworn volume(mL/mg) |
|-------|------------------|-------------------------|--------------------------|------------------------|---------------------|
| G–10 | <700 | 1.0±0.1 | 3 | 1 | 2–3 |
| G–15 | <1,500 | 1.5±0.2 | 3 | 1 | 2.5–3.5 |
| G–25 | <5,000 | 2.5±0.2 | 3 | 1 | 4–6 |
| G–50 | 1,500–20,000 | 8.0±0.3 | 3 | 1 | 9–11 |
| G–75 | 3,000–70,000 | 7.5±0.5 | 24 | 1 | 12–15 |
| G–100 | 4,000–150,000 | 10.0±1.0 | 72 | 1 | 15–20 |
| G–150 | 5,000–300,000 | 15.0±1.5 | 72 | 1 | 20–30 |
| G–200 | 5,000–600,000 | 20.0±2.0 | 72 | 1 | 30–40 |

The particle size of the gel can also affect the separation effect. The separation is better while using a gel with finer particle size, but with higher resistance and slower flow rate. Sephadex G–200 with 100–200 sieves is generally the better choice for the separation of proteins in the laboratory. Coarse–grained Sephadex G–25 and G–50 are usually used for desalting and preceded faster.

## 5.2.2   The gel pretreatment

The commercial gel is a dry pellet. It is necessary to be swelled directly in the eluent to be used. In order to accelerate the expansion, the mixture can be heated gradually in boiling water for 1–2 h, which can greatly accelerate the expansion. Especially when using a soft gel, it should take 24 h to several days to swell naturally, but can be completed within a few hours by heating. Besides saving more time, using this method can also remove the contaminated germs and remove the air from the gel.

## 5.2.3   Column packing and balancing

(1) Column selection

The column selection should follow similar rules in general column chromatography. The resolution depends on the gel height in the column and related to the square root of the gel height. However, the soft gel column will be deformed and obstructed if the gel is too high, and usually no more than 1 m. Thus, the gel height must be in a suitable range in order to separate the different components. A short column with a height of 20−30 cm is usually used in group separation. The ratio of the column height to diameter is (5−10) : 1, and the gel volume is 4−10 times of the sample solution volume. In fractionation separation, the ratio of the column height to diameter is (20−100) : 1.

The dead volume under the column filter pad should be as small as possible. The larger dead volume can more possibly cause the separated components being remixed. This can also affect the elution peak shape, lead to a trailing phenomenon, and decrease the resolution. In the accurate separation, the dead volume cannot exceed 1/1,000 of the gel bed volume. Based on the selected column, the column volume can be calculated via the formula $Vi = \pi d^2 h/4$, which is divided by the expansion coefficient of the gel and then can obtain the amount of the required dry gel.

(2) Column packing

Column packing is the key step for the gel filtration chromatography. Firstly, adding eluent in the column for about 1/3 height, the gel suspension is continuously and evenly poured into the column with stirring. Secondly, opening the column outlet while the gel deposited for 1−2 cm on the bottom of the column. Then, keep the solution flowing out the column and continuing to pour the gel suspension into the column. In the end, allowing the gel particles to settle slowly until the top surface of the gel is 3−5 cm from the top of the chromatography tube.

(3) Column balancing

After the column packing step, the elution buffer is continually added and equilibrate the gel column. Generally, the balancing process can be completed by using a buffer of 3−5 times the volume of the column bed to flow through the gel column at constant pressure. Checking the column gel, if it is uneven or appearing bubbles, the gel column needs to be refilled.

## 5.2.4   Sample loading and elution

(1) The treatment of sample solution

The precipitate of the sample solution should be filtered or centrifuged to be removed. The lipid impurities can be removed via rapid centrifugation or passing through a Sephadex (G−15) column. The sample viscosity can highly affect the separation efficiency. Thus, protein content should not exceed 40 g/L (4%). The volume of the sample should be controlled by the gel bed volume. The volume of protein sample to be separated is 1%−4% of the gel bed volume, and the sample volume can be 10% of the gel bed in the group separation. For protein desalting process, the sample solution volume can be up to 20%−30% of the gel bed. In fractionation separation, the sample volume should be very small, keeping the sample layer as thin as possible.

(2) Sample loading

The sample loading process is also a key step in gel chromatography. Before loading the sample, carefully check whether the column bed surface is flat enough. If it is uneven, gently stirring the surface with a thin glass rod to allow the gel to re−settle naturally and achieve a smooth surface. Close the outlet of the column just before the buffer on the column descends to the surface of the gel bed. The sample solution can be loaded carefully onto the gel bed by dropwise. Then, open the outlet of the column and adjust the appropriate flow rate until the sample solution penetrates in the gel bed. Carefully wash the inner wall of the column with the eluent to make sure the sample adhered to the column wall also penetrating into the gel.

(3) Elution

When the samples penetrate the gel bed, elution is continued with a certain buffer. The eluent should be the same as the solution soaked in the gel. Organic solvents (such as benzene and acetone) are usually used for water-insoluble samples separation; buffers with different ionic strength and pH value are commonly used for water-soluble sample separation. In addition, acidic eluents are for alkaline substances and alkaline eluents are for acidic substances.

## 5.2.5  Gel storage

The gel can be stored in two ways: the first one is keeping wet state and suitable for the commonly used gel, and the proper bacteriostat is needed; the second one is in the dry statement, which is suitable for the long-term preservation. Enzymes secreted by microorganisms can hydrolyze the glycoside bonds of polysaccharides in Sephadex and agarose, which can highly affect the gel properties as well as the separation efficiency. Thus, it is of significance to inhibit the microbial growth in certain gels.

There are three commonly used bacteriostatic which are listed as followings.

(1) Sodium azide ($NaN_3$): the little amount of the sodium azide (3 mmol/L, 0.02%) is sufficient to prevent the growth of microorganisms in gel filtration chromatography. However, sodium azide can interfere or even react with fluorescent-labeled proteins.

(2) Chloretone[$C_4H_7Cl_3O$]: it is usually used with 0.6-1.1 mmol/L (0.01%-0.02%) in the gel chromatography. It is better to be used in the weakly acidic solution and will decompose in a strongly alkaline solution or at the temperature higher than 60 ℃.

(3) Thimerosal: the commonly used concentration of the Thimerosal in gel chromatography is 0.12-0.5 mmol/L (0.005%-0.01%). It works well in weakly acidic solutions, and heavy metal ions should be avoided. In addition, it can be combined with the thiol group, and therefore, proteins containing thiol groups can reduce its antibacterial effect.

# 5.3    The application of gel chromatography

(1) For the separation and purification of macromolecules including protein, nucleic acid and so on.

(2) Desalting: in the separation of biochemical samples, various electrolytes could be introduced into the sample, while it is always necessary to add buffers with different pH value, or various salt for salting-out. Therefore, gel filtration can be used for desalting and leave the salt on the top of the gel. This method is very rapid and without changing the activity of the protein and enzyme.

(3) Concentrating: by utilizing the expedition property of the gel, the dry gel is added to the polymer solution to absorb water as well as small molecular substances. However, the polymer substance is blocked and remained in the external solution. After centrifugation or vacuum filtration, the solution and the expanded gel particles can be separated, and then the polymer material can be concentrated.

(4) For separation and refining: many substances, such as antibiotics, hormones, proteins, peptides, amino acids, vitamins, and alkaloids, can be purified by size exclusion chromatography. Molecular exclusion chromatography can also remove the pyrogen in pharmaceuticals and is also a good way to decolorize.

(5) Determination of relative molecular mass: the elution volume of globular proteins (relative molecular mass between 3,500 and 820,000) is mainly determined by its relative molecular mass. In the range of 3,500-820,000 of the relative molecular mass, the elution volume is approximately a linear function of the logarithm of the relative molecular mass. Therefore, using a calibration curve with the known relative molecular mass of the similar shape protein, the relative molecular mass of the other proteins can be estimated.

# Section 6   Affinity Chromatography

The method of separation and purification of biological macromolecules has been rapidly developed. Neutral salting−out or organic solvent precipitation, ion−exchange chromatography, gel filtration chromatography, electrophoresis, and other methods are commonly used. These methods either utilize the different solubility of the biopolymer, or by using differences in charge distribution and total charge in the molecule, or by using molecular size and shape differences. However, biopolymer content in tissue homogenates is usually very low and with a high level of impurities. In addition, the biomolecular do not have a significant difference in physicochemical properties. Therefore, it is of great difficulty to purify a certain biomolecular from the tissue.

## 6.1   The principle of affinity chromatography

It is known that bio−macromolecules can combine with certain compounds via reversible noncovalent bind, which is also called an affinity effect. Moreover, this combination is always specificity. For example, the active center of an enzyme can bind to a specific substrate, inhibitor, and cofactor sub bonds, and can be dissociated under certain conditions. The interaction of antigen and antibodies, hormones and receptors, ribonucleic acid and its complementary deoxyribonucleic acid also have similar characteristics. The ability of such polymers to form specific dissociable complexes with ligands is called binding affinity, and the substance reversibly bonded to the biomolecule is also called ligand. The chromatography developed based on this kind of reversible highly specific interaction is called affinity chromatography. Affinity chromatography is currently mainly used for protein purification. The target protein usually can be purified from a complex protein mixture in one step, and with high purity.

## 6.2   The basic process of affinity chromatography

The basic process of affinity chromatography is as follows.

(1) Combining the specific ligand with the carrier without affecting its biological function, known as immobilization or immobilization.

(2) Loading the carrier combined with the ligand to the column, that is called affinity chromatography column.

(3) Under certain conditions that favor the formation of a complex between the ligand and the biopolymer, the biopolymer mixture solution was added to the affinity column. The specific biomolecule of the mixed solution is adsorbed, and other impurities that are not adsorbed directly flow out. The column is washed thoroughly with a buffer to remove all non−adsorbed impurities.

(4) The purified biomolecule can be obtained by changing the solution that can dissociate the adsorbed biomolecule with the ligand.

(5) After fully washing the stationary phase, the affinity column can be regenerated and used for the next round of purification.

# 6.3    Selection and coupling of the carrier and ligand

## 6.3.1    The selection of carrier and ligand

The selection of affinity chromatography carrier must meet some basic rules.

(1) Being highly hydrophilic to make sure the ligands supported on the stationary phase be able to interact with the biomolecule in aqueous solution.

(2) Being inert to decrease the non-specific adsorption as lower as possible.

(3) Containing a large number of activated chemical groups, which can bind with the certain ligand under mild conditions.

(4) Being stable in physical, chemical and mechanical properties with a loose internal structure.

The commonly used stationary carrier in the affinity chromatography can be divided into two categories according to the chemical nature, that are a polysaccharide (such as agarose gel and Sephadex) and polyacrylamide. Agarose gel microspheres are the most commonly used, and particularly the type 4B (4% agarose gel) as well as type 6B and 2B.

The ideal ligand must own the high affinity for the biomolecule to be purified. In addition, suitable chemical group (such as amino group) should be contained to bind to the stationary carrier, but no affecting the specific binding of the ligand to the biomolecule.

## 6.3.2    Carriers and ligands coupling

There are many ways to couple ligands with the carriers, including the following four kinds. ① Adsorption: making the ligand to be adsorbed on the stationary carrier or ion exchanger. ② Covalent coupling: combining the ligand to the stationary via the chemical covalent bonds. ③ Cross-linking: exploring functional reagents to generate a reticular structure via cross-linking between the molecules. ④ Embedding: the ligand is wrapped in gel or polymer semipermeable membrane microcapsules.

The stationary carrier needs to be activated with appropriate reagents in order to combine the ligand with the carrier via stable covalent bonds. The reversible specific binding of ligand and biomolecule should be avoided. Two covalent coupling methods are listed as follows.

For the polysaccharide: the cyanide bromide (CNBr) is most commonly used, and periodates oxidation is another one. After treatment by these methods, the cyano group on the polysaccharide can be activated, and further, rapidly react with the amino groups of the proteins or other groups to form stable covalent bonds (Figure 5.8).

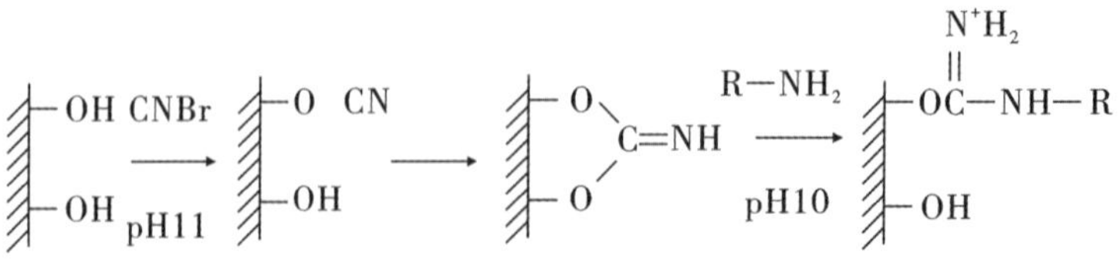

Figure 5.8    The cyanide bromide activation

Similar to the amino compound, the amide group on polyacrylamide can be stably combined with an aldehyde. Exploring glutaraldehyde, as a difunctional compound, can easily cross-link the polyacrylamide with protein or other compound containing amino groups via the aldehyde group. There are many kinds of activation reagents, such as carbodiimide compounds, succinic anhydride and so on.

Notably, the binding between the carrier and the ligand can hinder the reversible binding between the ligand and the biomolecule due to the surface of the ligand molecule being partly occupied, and resulting in so–called ineffective adsorption (steric hindrance). In this case, an appropriate linker can be attached between the carrier and the ligand, and greatly decrease the steric hindrance and effectively enhance the specific binding affinity (Figure 5.9). Both ethylenediamine and $\omega$–aminocaproic acid are good linkers. They act as arms via their active groups (amino, carboxyl, etc.) at both ends of the molecule, respectively.

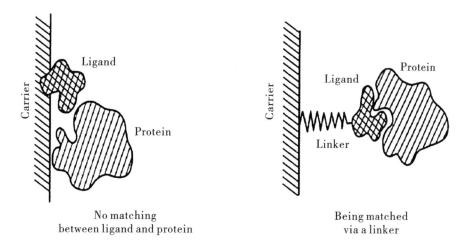

Figure 5.9    The binding affinity between ligand and protein affected by a linker

*Lin Li*

# Chapter 6

# Centrifugation

Centrifugation is a process that separates particles of interests from samples based on the physical principles of centrifugal force and particle settling. Centrifugal force makes suspension separated under high-speed rotation, and small particles (organelles, macromolecule precipitates, etc.) are settled at a certain rate. Centrifugation is an important experimental technique widely used in the life sciences, especially in biochemistry and molecular biology. It is mainly applied in the separation and preparation of various biological samples.

## Section 1   The Basic Principle of Centrifugation

Centrifugal force is generated when solid particles are moved circularly in a solution. The centrifugal force ($F$) can be expressed as the following formula:

$$F = ma = m\omega^2 r$$

"$F$" means the strength of the centrifugal force, "$a$" means the acceleration of the particle rotation, "$m$" means the effective mass of the settling particles, "$\omega$" means the angular velocity at which the particle is rotated, and "$r$" means the radius of rotation of the particle (cm).

Centrifugal force is usually expressed as multiples of the earth's gravitation (by the number $\times g$) and is called relative centrifugal force (RCF). For example 10,000 $\times g$, which means the relative centrifugal force of 10,000. Relative centrifugal force refers to the centrifugal force in a centrifugal field acting on the particles, which is equivalent to the multiples of earth's gravity. The unit is the acceleration of gravity, $g$ (980 cm/s$^2$).

$$RCF = ma/mg = m\omega^2 r/mg = \omega^2 r/g$$
$$\omega = 2\pi \times rpm/60$$
$$RCF = 1.119 \times 10^5 \times (rpm)^2 r$$

The rpm means revolutions per minute, which can also be expressed as r/min. It can be seen from the above equation that, as long as we know the radius of gyration $r$, RCF and rpm can be converted between each other. However, due to the difference in shapes and structures of rotors, centrifuge tubes from each centrifuge have different distances between every point from the nozzle to the bottom of the tube and the rotation axis. Therefore, in the calculation, the specified radius of rotation is replaced by the average radius $r_{av}$:

$$r_{av} = (r_{min} + r_{max})/2$$

Usually, "rpm" is used to describe low-speed centrifugation, meanwhile "$g$" is used to describe high-speed centrifugation. It's more scientific to use "RCF" than "rpm" in describing centrifugation conditions.

Because it can precisely reflect centrifugal forces and their dynamics change of particles in different locations within centrifuge tubes. we use RCF average to representative centrifugal force data in the scientific literature.

# Section 2  Type of Centrifuge

The centrifuge is a device that, implements centrifugation. According to the purpose of centrifugation, centrifuges can be divided into two categories, preparative centrifuge, and analytical centrifuge. The former is mainly used for the separation of biological materials, and its' sample capacity is relatively large. The latter is mainly used in studying the properties of pure macromolecules, including some granules, such as nucleoprotein and other substances, and the sample capacity to be analyzed is very small. Depending on the behavior of the substance to be examined in the centrifugal field (which can be continuously monitored by the optical system in the centrifuge), we can deduce its purity, shape, relative molecular mass and so on. Because they serve different purposes, the two types of centrifuges structurally different from each other. According to the speed range of the rotor, centrifuges can also be divided into three categories, ordinary centrifuge, high-speed centrifuge, and ultracentrifuge. Almost all analytical centrifuges are ultracentrifuges.

## 2.1  Preparative centrifuge

The preparative centrifuge can be divided into three categories as follows.

### 2.1.1  Ordinary centrifuge

An ordinary centrifuge's highest speed is approximately 6,000 r/min, and the maximum relative centrifugal force is around 6,000×g. The capacity of centrifuge tubes differs from milliliters to several liters. The form of separation is solid-liquid sedimentation separation. However, an ordinary centrifuge usually operates at room temperature. and it's speed cannot be strictly controlled without a refrigeration system. This type of centrifuge is commonly used to quickly collect and precipitated substances, such as red blood cells, crude sediment, and so on.

### 2.1.2  High-speed centrifuge

The maximum speed of a high-speed centrifuge is 20,000-25,000 r/min. The maximum relative centrifugal force is 89,000×g, while its' maximum capacity is 3 L. like the ordinary centrifuge, the high-speed centrifuge's form of separation is solid-liquid sedimentation separation. Furthermore, these centrifuges are equipped with freezers, which can eliminate heat generated by friction between rotor and air during high-speed rotation. The temperature, time and speed at centrifugation can be strictly controlled, which can be shown with a pointer or numerical display. In order to meet the needs of various preparations, different sizes of centrifuge tubes and rotors, are commonly used in the collection of microorganisms, cell debris, cells, larger organelles, ammonium sulfate precipitation, and immune precipitates.

### 2.1.3  Ultracentrifuge

The speed of an ultracentrifuge can reach 50,000-80,000 r/min, and the maximum relative centrifugal force can be 510,000×g. The sample volume can be up to 2 L. Based on the separation form, ultracentrifuge can be divided into differential centrifugation and density gradient centrifugation. The allowable error of a centrifuge tube is no more than 0.1 g. The advent of ultracentrifugation has brought a new development in the field of life science. It enables fractionated separation of subcellular organelles, and separation of macromolecules such as viruses, proteins, nucleic acids, and polysaccharides. Ultracentrifuges are widely used in the field of biochemistry and molecular biology.

# 2.2  Analytical centrifuge

Unlike the preparative centrifuge, the analytical centrifuge is applied in researches mainly in research on the settling characteristics and structures of biological macromolecules, rather than collecting a specific component. The analytical centrifuge uses a special rotating head and optical detection system so as to continuously monitor the settling process of material in a centrifugal field, thus determining the material's physical properties.

Much information can be obtained from a limited sample amount by using analytical centrifuges. For example, an analytical centrifuge makes it possible to determine whether a biological macromolecule exists or not and the approximation of its content. Furthermore, it can calculate the sedimentation coefficient of biological macromolecules, estimate the size of the molecules according to the interface diffusion, and detect the heterogeneity of the molecules and the proportion of components in the mixture. Additionally, it can also be used in detecting the conformation changes of biological macromolecules.

# Section 3    Separation Method of Preparative Centrifuge

# 3.1  Differential centrifugation

Differential centrifugation is one of the most common methods. It uses a gradual increase in centrifugal speed, which separates into batches the sedimentation of different particles separated in batches at different centrifugal speeds and times. This method is routinely used in separating larger particles by the sedimentation coefficient difference.

The centrifugal force and centrifugal time needed for particle settlement should be selected based on the settling characteristics of all particles to be separated. After the first round of centrifugation, the largest and heaviest particles can be precipitated at the bottom of the centrifuge tube. The separated supernatant can be transferred to a new centrifuge tube and be centrifuged at an increased speed to yield a second, larger, heavier pellet and supernatant containing smaller and lighter pellets. With many centrifugations, different particles are separated step by step. The sediment obtained by this method is heterogeneous, and the precipitated components always have contaminations of non−precipitated components. After 2−3 rounds of resuspension and re−centrifugation, the relatively pure particles can be obtained ( Figure 6.1).

This method is mainly used for the separation of organelles and viruses in tissue homogenates. The advantages of this method are the simplicity of operation, separating the supernatant and sediment by simply pouring the supernatant out, and using larger capacity angular rotors. The disadvantages of this method are that it can't be pure particles at one time, multiple rounds of centrifugation are needed, other impurities exist in the precipitate, the separation effect is limited, and the sample can be easily degenerated and inactivated.

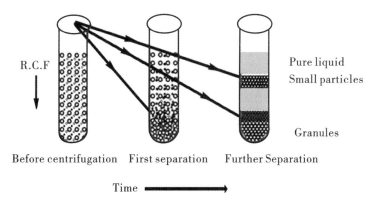

Figure 6.1    Differential centrifugation sketch

# 3.2    Density gradient centrifugation

Density gradient centrifugation is also referred to as zonal centrifugation. The sample to be separated is added to an inert gradient medium for centrifugal settlement or settlement equilibrium. Therefore, under certain centrifugal force, particles distribute to specific locations in the gradient matrix, forming different zones. The advantages of this method are as follows: ① the separation effect is satisfying, and the pure particles can be obtained with just once centrifugation; ② the application range is wide, this method can facilitate separating of particles with a large difference in sedimentation coefficient; ③particles maintain their activity, without crushing and deformation. The disadvantages of this method are the centrifugation takes a longer time, need to be prepared, and it requires a rigorous operation that is not easy to grasp.

Density gradient zone centrifugation can be divided into two types.

## 3.2.1    Zonal centrifugation

Particles with different sedimentation coefficients are settling at different speed under a certain centrifugal force so that the density gradient medium forms zones. This method is only used for the separation of particles with a more than 20% difference of sedimentation coefficient, regardless of the density. Particles of the same size but with different densities ( such as mitochondria, lysosomes, etc. ) cannot be separated by this method. The key to this centrifugation method is to select the appropriate centrifugal speed and time.

Centrifuge tube installed density gradient medium, the sample solutionis added at the top of the gradient medium. Due to the centrifugal force, particles leave the original sample layer and settle to the bottom according to their settlement rate. After centrifugation, the settled particles form a series of clearly discontinuous zones ( Figure 6.2). The larger the settlement factor, the faster the particle settles, and the lower the zone is presented. Centrifugation must be ended before the fastest particles reach the bottom of the tube. In zonal centrifugation, the density of sample particles is higher than the density of the gradient medium. Usually, a sucrose solution is applied as the gradient medium, with a maximum density of 1.28 $kg/m^3$and with a concentration of up to 60% . For the separation of living cells, such as isolation and purification of primary cells or lymphocytes, high density of small molecule substances can interfere with cell survival. Synthetic hydrophilic polymers are used as density gradient mediums. Ficoll and Percoll are two kinds of commonly used hydrophilic polymer macromolecules.

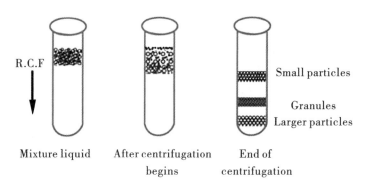

Figure 6.2    Differential zone centrifugation sketch

## 3.2.2    Equilibrium centrifugation

There are two kinds of equilibrium centrifugation methods to generate gradient: pre-formed gradient and centrifugal gradient. In the former method, the pre-gradient medium is applied in a centrifuge tube, and the sample is added to the surface of the gradient matrix. In the latter method, samples are pre-mixed with the gradient medium, and the gradient can be formed via centrifugation.

During centrifugation, different particles move to a specific gradient layer according to their densities and congregate in different zones. Once the system reaches equilibrium, it is meaningless to extend the centrifugal time and increase rotational speed. The shapes and positions of zones formed by particles at the isopycnic point are no longer affected by the centrifugal time. However, increasing the rotational speed can shorten the centrifugal time needed for reaching equilibrium. The time required for centrifugation usually can be up to several days, based on the time it takes for the smallest particles to reach the point of equal density ( in the equilibrium point).

The separation efficiency of the equilibrium centrifugation depends on the buoyant-density difference of sample particles. The greater the density difference, the better the separation, regardless of the particles' sizes and shapes. However particle sizes and shapes determine the centrifugal time needed for reaching equilibrium, and the width of the zone formed. The gradient medium used for equilibrium centrifugation is usually chlorinated ( CsCl), with a density of 1. 7 $g/cm^3$. This method separates nucleic acids, subcellular organelles, etc. , including complex proteins, but it is not suitable for simple proteins( Figure 6. 3) .

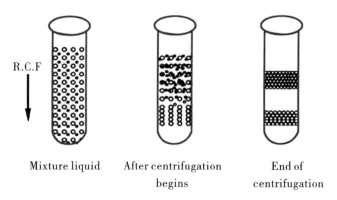

Figure 6.3    Equal density zone centrifugation sketch

There are many ways to collect separated zones, including: ①aspirated from the upper part of the centrifuge tubes with a syringe and eyedropper; ②dripped through the bottom of the centrifuge tube with a needle piercing; ③piercing the wall of the zone section of the centrifuge tube with a needle and pulling out the sample zone; ④ a thin tube into the bottom of the centrifuge tube so that the replacement liquid with the maximum density of the gradient medium can be pumped. Then, samples and the gradient medium are pressed out and collected by an automatic collector.

# Section 4   Operation Guides and Precautions of Centrifugation

An ultracentrifuge is important precise equipment in teaching and research of biochemistry and molecular biology experiments. Because of its high speed and large centrifugal force, serious accidents may occur due to improper usage or lack of regular maintenance and repair. Therefore, the following operation guides and precautions should be strictly obeyed.

(1) Centrifuge tubes and their contents must be carefully balanced. The allowable difference between balanced centrifuge tubes should not exceed the range mentioned in the instructions of each centrifuge. Each centrifuge has its own allowable difference with the different rotors. An odd number of tubes cannot be loaded in the rotor. When the rotor is only partially loaded, the tubes must be placed symmetrically so that the load is evenly distributed.

(2) When the solution is loaded in centrifugation tubes, a suitable centrifuge tube must be selected according to the nature and volume of the centrifugal liquid. In the case of non-cover centrifuge tubes, the liquid must not be overloaded in order to prevent the centrifuge from being flung out, resulting in unbalance, rust or corrosion. When using the centrifuge tube of the preparation ultracentrifuge, it is often required that the liquid must be filled to avoid denting the upper portion of the plastic centrifuge tube during centrifugation. It is strictly prohibited to use a centrifuge tube with significant deformation, damage or aging. It must be carefully checked, and promptly cleaned and dried after usage. The rotor is one of the most important parts in centrifuges and should be carefully protected. The rotor must be moved carefully to avoid colliding and injuries. A layer of glazed wax should be applied to protect the rotor if it is not needed for a long time.

(3) If centrifugation occurs at a temperature below room temperature, the rotor should be placed in the refrigerator or in the centrifuge rotor precooling system before using.

(4) During centrifugation, the operator is not allowed to leave at any time. Continuous observation is required to make sure the centrifuge is working properly normally. Once any abnormal sound occurs, the centrifuge should be shut down immediately to check for and troubleshoot any problems.

(5) Each rotor has its own maximum allowable speed. Before using the rotor, the instructions should be consulted. Each rotor should be equipped with a usage note, and cumulative usages should be recorded.

*Zhang Yuzhe, Hu Xiaojuan*

# Part II

# Protein Experiments

# Chapter 7

# Quantitative Analysis of Protein by Spectrophotometry

Determination of protein concentration ( quantitative analysis of protein) is very necessary for many kinds of protein-related studies. There are many methods to measure protein concentration, such as Biuret assay, Lowry assay, BCA assay, Bradford assay, and spectrophotometry based on UV absorption. Biuret assay is simple and easy to operate with no need for expensive instruments, which is often used in laboratory decades ago. Lowry assay is developed from Biuret assay and its sensitivity is increased 100 folds. BCA is also developed from Biuret assay and is the commonly used method in research at present due to its high sensitivity and trace detection. Every method has its advantages and disadvantages ( Table 7.1). There is no ideal method to measure protein concentration. In practice, the selection of protein assay method is usually based on amino acid composition, the structure of the protein, sample treatment, experimental condition and available instruments. Some criteria for choosing an assay include: ① the compatibility of reagents with protein sample; ② assay uniformity between different proteins; ③ operability and assay time required; ④ sample volume and concentration range; ⑤ performance and assay range of the spectrophotometer.

Table 7.1　Methods of protein concentration determination

| Method | Sensitivity ( μg/mL) | Advantages | Disadvantages |
|---|---|---|---|
| Biuret assay | 1,000–20,000 | Quick, less interfering substances | Low sensitivity |
| Lowry assay | 20–250 | High sensitivity | Much interfering substances, time-consuming |
| BCA assay | 20–2,000 | Highly sensitive, less interfering substances | Interfered by urea and $\beta$-mercaptoethanol |
| Bradford assay | 10–1,000 | Simple, quick and high sensitivity | Large variation between different proteins |
| UV spectro-photometry | 100–1,000 | Rapid, easy, recoverable | Relies on aromatic residues |

# Experiment 1    Determination of Protein Concentration by Biuret Assay

## 1.1    Principle

The biuret reaction can be used for both qualitative and quantitative analysis of protein. Two molecules of urea can be condensed to form biuret ($H_2NOC$—$NH$—$CONH_2$) containing amido bond under high temperature and one molecule of $NH_3$ is released. In a moderately alkaline medium, the nitrogen of the amido bond can react with cupric ions ($Cu^{2+}$) to form a purple-colored complex, which is a colorimetric biuret reaction.

The protein contains many peptide bonds which are similar to amido bonds in biuret and can produce $Cu^{2+}$ – peptide compound in biuret reaction (Figure 7.1). Within a certain concentration range, the intensity of the color is proportional to the number of peptide bonds that are reacting, and therefore is directly proportional to the protein concentration. The reaction doesn't occur with free amino acids because of the absence of peptide bonds. The method is easy to operate and quick to obtain the result. Different kinds of amino acids in the primary structure of protein do not affect the test result. But Biuret assay is not very sensitive (1 –20 mg protein) comparing with other methods and the reaction also occurs with any compound containing the bonds, such as —HN—CO—, —HN—CH$_2$— and —HN—CS—.

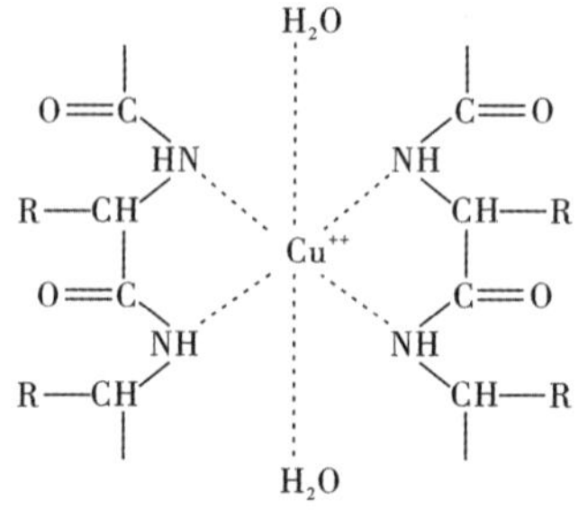

Figure 7.1    $Cu^{2+}$ – peptide complex

## 1.2    Materials

### 1.2.1    Reagents and solutions

(1) Sample: unknown protein sample or serum.

(2) Biuret reagent (1 L): 1.5 g $CuSO_4 \cdot 5H_2O$, 6.0 g potassium sodium tartrate, 300 mL of 10% NaOH, 1.0 g KI, qs to 1,000 mL with Milli-Q $H_2O$.

Put the solution in a brown bottle and if dark red deposition appears, the reagent cannot be used.

(3) 2 mg/mL BSA solution (100 mL): 0.2 g BSA qs to 100 mL with Milli-Q $H_2O$.

### 1.2.2    Special equipment

①Spectrophotometer; ②micropipettes and tubes; ③thermostat water bath.

# 1.3    Methods

## 1.3.1    Prepare the standard curve

(1) Label 7 test tubes as 0–6 and place them in a test tube rack.

(2) Add the solutions in the following Table 7.2 to each tube.

Table 7.2    Preparation of Biuret standards

| Reagents (mL) | 0 | 1 | 2 | 3 | 4 | 5 | 6 |
|---|---|---|---|---|---|---|---|
| 2 mg/mL BSA | — | 0.3 | 0.6 | 1.2 | 1.8 | 2.4 | 3.0 |
| Distilled water | 3.0 | 2.7 | 2.4 | 1.8 | 1.2 | 0.6 | — |
| BSA concentration (mg/mL) | 0 | 0.2 | 0.4 | 0.8 | 1.2 | 1.6 | 2.0 |

(3) Add 3.0 mL Biuret reagent to each tube.

(4) Mix well by vortex mixer and incubate at 37 ℃ for 30 min.

(5) Read the absorbance for each tube against the blank (number 0 tube in Table 7.2) at 540 nm.

In order to guarantee the accuracy of detection, each concentration should be duplicate or triplicate and use the mean of parallel absorbance.

## 1.3.2    Determine the sample

Add 3.0 mL unknown protein sample in a tube and add 3.0 mL of Biuret reagent. Then mix well and incubate at 37 ℃ for 30 min. Finally, measure the absorbance at 540 nm. Procedure 1 and 2 can be operated together.

# 1.4    Results and discussion

## 1.4.1    Plot the standard curve

According to the testing result of Table 7.3, plot a standard curve by using absorbance ($A_{540}$) as $y$–axis and BSA concentration as $x$–axis, which can be drawn in EXCEL software. A standard curve should be a linear line across the origin $(0,0)$ (Figure 7.2) and a formula is also obtained. The values determined can be variable because of a different batch of reagents or different operation. Use the real data obtained to plot the standard curve.

Table 7.3    The absorbance of different protein standard solution

| Protein standard concentration (mg/mL) | 0 | 0.2 | 0.4 | 0.8 | 1.2 | 1.6 | 2.0 |
|---|---|---|---|---|---|---|---|
| Absorbance | 0 | 0.037 | 0.063 | 0.096 | 0.150 | 0.197 | 0.251 |

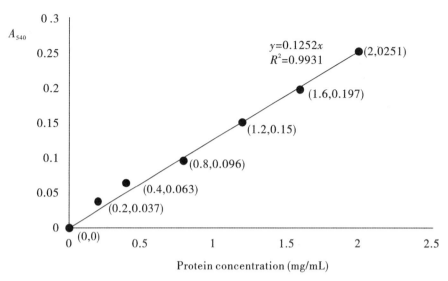

Figure 7.2  A standard curve for Biuret assay

## 1.4.2  Calculate the concentration of unknown protein sample

(1) Standard curve method: protein concentration can be calculated according to the formula obtained from the standard curve by replacing $y$ as $A_{540}$ of the unknown sample. In this example, the unknown sample concentration can be calculated by the formula of $y=0.1252x$ (Figure 7.2).

(2) Standard addition method: the unknown sample (u) can also be determined directly by comparing the $A_u$ of the unknown sample to the $A_s$ of a standard solution whose concentration is known to provide that such compounds obey the Lambert–Beer law and all conditions under which standards and unknowns prepared should be kept identical.

Standard sample $A_s = K_s \times C_s \times L_s$

Unknown sample $A_u = K_u \times C_u \times L_u$

Since the $L$, the path length and the extinction coefficient will be constant; that is, $L_s = L_u$ and $K_s = K_u$, thus, $A_s/A_u = C_s/C_u$, $C_u = A_u/A_s \times C_s$.

## 1.5  Troubleshooting

(1) Biuret assay is not so sensitive to determine protein as low as 1 mg. When protein content in the sample is low, it is better to select another method with much more sensitivity.

(2) The standard curve should be a straight line through the origin. When protein concentration is 0, the absorbance is 0. So intercept (0,0) is selected when plotting.

(3) $R^2$ in Figure 7.2 shows the linear regression correlation between two variables $(x, y)$. The more $R^2$ closer to 1 indicates the higher correlation between two variables. In this experiment, good data can make $R^2$ more than 0.99. The low $R^2$ may be related to the inaccuracy of sample adding.

(4) The assay of the unknown sample should be within the test range of the standard curve and if it exceeds the test range, the unknown sample should be diluted or concentrated in a certain proportion before determination. Do not forget to multiply the folds of dilution when calculating the final concentration of the unknown sample.

# Experiment 2   Determination of Protein Concentration by Lowry assay

## 2.1   Principle

Lowry protein assay is a method developed from Biuret assay by adding another Folin–phenol reagent (a mixture of phosphotungstic acid and phosphomolybdic acid). The method combines Biuret assay with the oxidation of aromatic amino acid residues of the protein. In the first reaction, $Cu^{2+}$ react with protein to form $Cu^{2+}$–protein complex and is reduced to $Cu^+$ by peptide bond under alkaline condition. In the second reaction, Folin–phenol reagent is reduced by aromatic amino acid residues of protein (mainly tryptophan and tyrosine) to form a blue complex (a mixture of tungsten blue and molybdenum blue). Within a certain concentration range, the intensity of the blue color is directly proportional to the protein concentration. Lowry method is much more sensitive and can determine protein as low as 5 $\mu g/mL$. But the different composition of tryptophan and tyrosine in the primary structure can affect the test result. In addition, Lowry reaction consumes much time and can be interfered by some chemicals such as citrate acid, phenol and so on.

## 2.2   Materials

### 2.2.1   Reagents and solutions

(1) Sample: unknown protein sample or serum.

(2) Solution A(500 mL): 10 g $Na_2CO_3$, 2 g NaOH, 0.25 g potassium sodium tartrate ($KNaC_4H_4O_6 \cdot 4H_2O$), qs to 500 mL with Milli–Q $H_2O$.

(3) Solution B(500 mL): 0.5 g copper sulfate($CuSO_4 \cdot 5H_2O$), qs to 100 mL with Milli–Q $H_2O$.

(4) Folin – phenol reagent: 100 g sodium tungstate ($Na_2WO_4 \cdot 2H_2O$), 25 g sodium molybdate ($Na_2MoO_4 \cdot 2H_2O$), 50 mL of 85% phosphoric acid, 100 mL concentrated hydrochloric acid, qs to 700 mL with Milli–Q $H_2O$. Then mix well and reflux for 10 h, add 150 g lithium sulfate($Li_2SO_4$), 50 mL distilled $H_2O$ and drops of bromine water. Continue boiling for 15 min with the flask mouth open in order to get rid of excessive bromine, and cool down to get a yellow solution (if it's still green, repeat adding bromine and boiling), qs to 1,000 mL with Milli–Q $H_2O$ and perform filtration.

The Folin–phenol reagent is stored in a brown bottle. A commercial reagent is recommended.

(5) 2 mg/mL BSA solution(100 mL): 0.2 g BSA qs to 100 mL with Milli–Q $H_2O$.

### 2.2.2   Special equipment

①Spectrophotometer; ②micropipettes and tubes; ③thermostat water bath.

## 2.3   Methods

### 2.3.1   Prepare the standard curve

(1) Prepare the reaction solution: the reaction solution is a mixture of solution A and B, in which solu-

tion A is mixed with solution B in the volume ratio of 50 : 1. According to the number of samples, prepare sufficient reaction solution and mix well.

Example: preparing 10 mL reaction solution, for easy calculation, take 10 mL solution A, 0.2 mL (10/50) solution B, mix well and ready for use.

(2) Determine protein standard solution

1) Label 7 test tubes and add the solutions in Table 7.4 to each tube.

Table 7.4　Standard curve plotting by Lowry assay

| Reagents (μL) | 0 | 1 | 2 | 3 | 4 | 5 | 6 |
| --- | --- | --- | --- | --- | --- | --- | --- |
| 2 mg/mL BSA | — | 2 | 4 | 6 | 8 | 10 | 12 |
| Distilled water | 200 | 198 | 196 | 194 | 192 | 190 | 188 |
| BSA concentration(μg/mL) | 0 | 20 | 40 | 60 | 80 | 100 | 120 |

2) Add 1.0 mL the reaction solution to each tube.

3) Mix well by vortex mixer and incubate at room temperature for 10 min.

4) Add 100 μL Folin-phenol reagent to each tube and incubate at room temperature for 30 min.

5) Read the absorbance for each tube against the blank (number 0 tube in Table 7.4) at 750 nm.

In order to guarantee the accuracy of detection, each concentration should be duplicate or triplicate and use the mean of parallel absorbance.

## 2.3.2　Determine the sample

Add 200 μL unknown protein sample in a new tube and perform colorimetric reaction as above step (2). Procedure 2.3.1 and 2.3.2 can be operated together.

# 2.4　Results and discussion

## 2.4.1　Plot the standard curve

According to the test result of Table 7.4, plot a standard curve by using absorbance ($A_{750}$) as $y$-axis and BSA concentration as $x$-axis, which can be drawn in EXCEL software. A standard curve should be a linear line across the origin (0,0) (refer to Figure 7.2) and the formula is also obtained. The values determined can be variable because of a different batch of reagents or different operation. Use the real data obtained to plot the standard curve.

## 2.4.2　Calculate the concentration of unknown protein sample

(1) Standard curve method: protein concentration can be calculated according to the formula from the standard curve by replacing $y$ as $A_{750}$ of the unknown sample.

(2) Standard addition method: $C_u = A_u/A_s \times C_s$.

$C_u$: concentration of the unknown sample; $C_s$: concentration of the standard solution.

$A_u$: absorbance of the unknown sample; $A_s$ absorbance of the standard solution.

## 2.5　Troubleshooting

(1) Reaction solution should be freshly prepared before use. It cannot be stored for a long time.

(2) Mix well quickly after Folin−phenol reagent is added in order to avoid the inactivation of phospho-tungstic acid and phosphomolybdic acid before they are reduced.

(3) The standard curve should be a straight line through the origin. When protein concentration is 0, the absorbance is 0. So intercept $(0,0)$ is selected when plotting.

(4) $R^2$ in Figure 7.2 shows the linear regression correlation between two variables $(x, y)$. The more $R^2$ closer to 1 indicates the higher correlation between two variables. In this experiment, good data can make $R^2$ more than 0.99. The low $R^2$ may be related to the inaccuracy of sample adding.

(5) The assay of the unknown sample should be within the test range of the standard curve and if it exceeds the test range, the unknown sample should be diluted in a certain proportion before determination. Do not forget to multiply the folds of dilution when calculating the final concentration of the unknown sample.

# Experiment 3　Determination of Protein Concentration by BCA Assay

## 3.1　Principle

The bicinchoninic acid (BCA) protein assay is a detergent−compatible formulation based on BCA for the colorimetric detection and quantitation of total protein. This method also combines the well−known reduction of cuprous cation $Cu^{2+}$ to $Cu^+$ by protein in an alkaline medium (the biuret reaction) with the highly sensitive and selective colorimetric detection of the $Cu^+$ using a unique reagent containing BCA. The purple−colored reaction product of this assay is formed by the chelation of two molecules of BCA with one $Cu^+$ (Figure 7.3). This water−soluble complex exhibits a strong absorbance at 562 nm that is nearly linear with increasing protein concentrations over a broad working range (0.5−2,000 μg/mL). The BCA assay is not a true end−point method, that is, the final color continues to develop. However, following incubation, the rate of continued color development is sufficiently slow to allow large numbers of samples to be assayed together. The number of peptide bonds and the presence of four particular amino acids (cysteine, cystine, tryptophan, and tyrosine) are reported to be responsible for color formation with BCA. Comparing with other methods, BCA assay is highly sensitive and has good compatibility with substances such as low concentration of SDS, Triton X−100 and Tween−20.

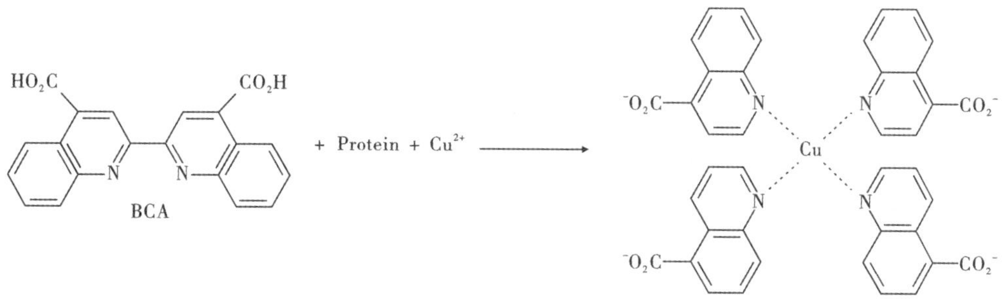

Figure 7.3　BCA · $Cu^+$ complex in BCA assay

## 3.2 Materials

### 3.2.1 Reagents and solutions

(1) Sample: unknown protein sample or serum.

(2) BCA reagent A: 1 g BCA, 2 g $Na_2CO_3$, 0. 16 g sodium tartrate, 0. 4 g NaOH, 0. 95 g $NaHCO_3$, qs to 100 mL with Milli-Q $H_2O$, adjust pH to be 11. 25.

(3) BCA reagent B: 0. 4 g $CuSO_4$, in 10 mL Milli-Q $H_2O$.

(4) 2 mg/mL BSA solution(100mL): 0. 2 g BSA qs to 100 mL with Milli-Q $H_2O$.

### 3.2.2 Special equipment

①Spectrophotometer; ②microplate; ③micropipettes and tubes; ④thermostat water bath.

## 3.3 Methods

### 3.3.1 Prepare the BCA working solution

BCA working solution is a mixture of solution A and B, in which solution A is mixed with solution B in the volume ration of 50 : 1. According to the number of samples, prepare sufficient reaction solution and mix well.

Example: prepare 10 mL working solution, for easy calculation, take 10 mL solution A, 0. 2 mL(10/50) solution B, mix well and ready for use.

### 3.3.2 Prepare the standard curve and determine the sample (microplate procedure)

(1) Label 9 test tubes and add the solutions in Table 7. 5 to each tube.

Table 7.5  **Preparation of BSA standard solutions**

| Vial | ddH$_2$O( μL) | 2 mg/mL BSA ( μL) final | BSA concentration ( mg/mL) |
|------|---------------|-------------------------|----------------------------|
| A | 0 | 300 of Standard | 2. 0 |
| B | 125 | 375 of Standard | 1. 5 |
| C | 325 | 325 of Standard | 1. 0 |
| D | 175 | 175 of vial B dilution | 0. 75 |
| E | 325 | 325 of vial C dilution | 0. 5 |
| F | 325 | 325 of vial E dilution | 0. 25 |
| G | 325 | 325 of vial F dilution | 0. 125 |
| H | 400 | 100 of vial G dilution | 0. 025 |
| I | 400 | 0 | 0 = Blank |

(2) Pipette 25 μL each standard and unknown sample into a microplate well.

If the sample volume is limited, the sample can be diluted before detection, for example, the 5 μL sample can be diluted with 20 μL ddH$_2$O.

(3) Add 200 μL the BCA working solution to each well and mix plate thoroughly on a plate shaker for 30 s.

(4) Cover the plate and incubate at 37 ℃ for 30 min.

(5) Cool plate to room temperature. Measure the absorbance at or near 562 nm on a plate reader.

(6) Subtract the average $A_{562}$ of the Blank ( Vial I, Table 7.5) from the $A_{562}$ of all other standards and sample.

In order to guarantee the accuracy of detection, each concentration should be duplicate or triplicate and use the mean of parallel absorbance.

# 3.4   Results and discussion

## 3.4.1   Plot the standard curve

According to the test result of Table 7.5, plot a standard curve by using absorbance ( $A_{562}$ ) as $y$−axis and BSA concentration as the $x$−axis, which can be drawn in EXCEL software. A standard curve should be a linear line across the origin (0,0) ( refer to Figure 7.2) and the formula is also obtained. The values determined can be variable because of a different batch of reagents or different operation. Use the real data obtained to plot the standard curve.

## 3.4.2   Calculate the concentration of unknown protein sample

(1) Standard curve method: protein concentration can be calculated according to the formula obtained from the standard curve by replacing y as $A_{562}$ of the unknown sample.

(2) Standard addition method: $C_u = A_u/A_s \times C_s$.

$C_u$: concentration of the unknown sample; $C_s$: concentration of the standard solution.

$A_u$: absorbance of the unknown sample; $A_s$: absorbance of the standard solution.

# 3.5   Troubleshooting

(1) Wavelengths from 540−590 nm have been used successfully with this method.

(2) BCA working solution should be freshly prepared before use and can not be stored for a long time.

(3) The standard curve should be a straight line through the origin. When protein concentration is 0, the absorbance is 0. So intercept (0,0) is selected when plotting.

(4) $R^2$ in Figure 7.2 shows the linear regression correlation between two variables ( $x$, $y$) . The more $R^2$ closer to 1 indicates the higher correlation between two variables. In this experiment, good data can make $R^2$ more than 0.99. The low $R^2$ may be related to the inaccuracy of sample adding.

(5) The assay of the unknown sample should be within the test range of the standard curve and if it exceeds the test range, the unknown sample should be diluted in a certain proportion before determination. Do not forget to multiply the folds of dilution when calculating the final concentration of the unknown sample.

# Experiment 4    Determination of Protein Concentration by Bradford Assay

## 4.1    Principle

The Bradford assay is a protein determination method that involves the binding of Coomassie Brilliant Blue G-250 dye to proteins. The dye exists in three forms: cationic (red), neutral (green), and anionic (blue). Under acidic conditions, the dye is predominantly in the doubly protonated red cationic form ($A_{max} = 470$ nm). However, when the dye binds to the protein, it is converted to a stable unprotonated blue form ($A_{max} = 595$ nm). It is this blue protein-dye form that is detected at 595 nm in the assay using a spectrophotometer or microplate reader. Coomassie Brilliant Blue G-250 dye binds primarily to basic (especially arginine) and aromatic amino acid residues. Certain chemical-protein and chemical-dye interactions interfere with the assay. Interference from non-protein compounds is due to their ability to shift the equilibrium levels of the dye among the three colored species. Known sources of interference, such as some detergents, flavonoids, and basic protein buffers, stabilize the green neutral dye species by direct binding or by shifting the pH. Nevertheless, many chemical reagents do not directly affect the development of dye color when used in the standard protocol. Bradford assay is sensitive and can detect protein concentration as low as 1 μg/mL.

## 4.2    Materials

### 4.2.1    Reagents and solutions

(1) Sample: unknown protein sample or serum.

(2) Coomassie Brilliant Blue G-250: dissolve 100 mg Coomassie Brilliant Blue G-250 in 50 mL 95% ethonal, then add 120 mL 85% phosphoric acid ($H_3PO_4$), qs to 1L ddH$_2$O.

(3) PBS(pH 7.2-7.4): NaCl 137 mmol/L, KCl 2.7 mmol/L, Na$_2$HPO$_4$ 10 mmol/L, KH$_2$PO$_4$ 1.76 mmol/L.

(4) 2 mg/mL BSA solution(100 mL): 0.2 g BSA qs to 100 mL with Milli-Q H$_2$O.

### 4.2.2    Special equipment

①Spectrophotometer; ②micropipettes and tubes; ③thermostat water bath.

## 4.3    Methods

(1) Label 6 test tubes and add the solutions in Table 7.6 to each tube.

Table 7.6    Preparation of BSA standards

| Reagent($\mu$L) | 0 | 1 | 2 | 3 | 4 | 5 |
|---|---|---|---|---|---|---|
| 2 mg/mL BSA | 0 | 5 | 10 | 20 | 30 | 60 |
| PBS | 150 | 145 | 140 | 130 | 120 | 90 |
| BSA conc. (mg/mL) | 0 | 0.067 | 0.13 | 0.27 | 0.4 | 0.8 |

(2) Add 2.85 mL Coomassie Blue G-250 to each tube and a tube with a 150 $\mu$L unknown protein sample, then mix by vortex or inversion.

If the sample volume is limited, the sample can be diluted before detection, for example, 20 $\mu$L sample can be diluted with 130 $\mu$L ddH$_2$O.

(3) Incubate at room temperature for at least 5 min. Samples should not be incubated longer than 1 h at room temperature.

(4) Set the wavelength of the spectrophotometer to 595 nm. Zero the instrument with the blank (0 tube) sample. Measure the absorbance of the standards and unknown sample.

In order to guarantee the accuracy of detection, each concentration should be duplicate or triplicate and use the mean of parallel absorbance.

# 4.4    Results and discussion

## 4.4.1    Plot the standard curve

According to the test result of table 7.6, plot a standard curve by using absorbance ($A_{595}$) as y-axis and BSA concentration as x-axis, which can be drawn in EXCEL software. A standard curve should be a linear line across the origin (0,0) (refer to Figure 7.2) and the formula is also obtained. The values determined can be variable because of a different batch of reagents or different operation. Use the real data obtained to plot the standard curve.

## 4.4.2    Calculate the concentration of unknown protein sample

(1) Standard curve method: protein concentration can be calculated according to the formula from the standard curve by replacing y as $A_{595}$ of the unknown sample.

(2) Standard addition method: $C_u = A_u/A_s \times C_s$.

$C_u$: concentration of the unknown sample; $C_s$: concentration of the standard solution.

$A_u$: absorbance of the unknown sample; $A_s$: absorbance of the standard solution.

# 4.5    Troubleshooting

(1) Glass or plastic cuvette should be used in the experiment because Coomassie Brilliant Blue G-250 can firmly bind on quartz cuvette.

(2) Much more accurate results can be obtained if the standard curve is created by analyzed protein (sample protein) because different proteins have the different binding capacity with Coomassie Brilliant Blue G due to different amino acids composition.

(3) The standard curve should be a straight line through the origin. When protein concentration is 0, the absorbance is 0. So intercept (0,0) is selected when plotting.

(4) $R^2$ in Figure 7.2 shows the linear regression correlation between two variables $(x, y)$. The more $R^2$ closer to 1 indicates the higher correlation between two variables. In this experiment, good data can make $R^2$ more than 0.99. The low $R^2$ may be related to the inaccuracy of sample adding.

(5) The assay of the unknown sample should be within the test range of the standard curve and if it exceeds the test range, the unknown sample should be diluted in a certain proportion before determination. Do not forget to multiply the folds of dilution when calculating the final concentration of the unknown sample.

*Li Jiao, Lü Lixia*

# Chapter 8

## Separation and Quantitative Analysis of Proteins by Electrophoresis

Electrophoresis is a powerful technique that is widely used to fractionate proteins. Electrophoresis depends on the ability of charged molecules to migrate when placed in an electric field.

## Experiment 1   Separation and Quantification of Serum Proteins by Cellulose Acetate Membrane Electrophoresis

## 1.1   Principle

Cellulose acetate is the acetate ester of cellulose produced from the reaction of cellulose with acetic anhydride. Protein samples were separated electrophoretically on a cellulose acetate membrane. Cellulose acetate membrane electrophoresis ( CAME) can be useful in identifying multimeric proteins formed by different subunits since each subunit has a different charge due to the different amino acid composition. CAME is an important technique in clinical diagnostics.

A mixture of proteins such as serum is dotted on the marked center of a cellulose acetate strip. The strip is then put across two tanks of the electrophoresis chamber, which are filled with barbital buffer. Desired pH and voltage are applied across the strip. The proteins that migrate towards the anode have a pI greater than the pH of the buffer, while proteins that migrate towards the cathode have a pI less than the pH of the buffer. Positively charged proteins migrate towards the cathode while negatively charged proteins migrate toward the anode. Individual molecules migrate differently so that separation bands can be observed.

Microliters of serum are loaded on the cellulose–acetate strip. After setting current, the serum proteins migrate on the membrane with different velocities towards the anode. When electrophoresis stops, different proteins are already located in different positions of the strips. The nearest band towards anode is albumin, followed by $\alpha_1$, $\alpha_2$, $\beta$, and $\gamma$–globulins. The strip is then removed from the chamber. After staining, several colored bands appear on the strip.

The concentration of total plasma proteins is 60–80 g/L. Albumin is the major protein of human plasma ( 38–48 g/L), nearly 50% of the total plasma proteins. The liver produces about 12 g of albumin per day. Albumin has two main functions. ① Maintaining colloid osmotic pressure of blood. ② Transportation: actin as a carrier molecule for bilirubin, fatty acids, trace elements, and many drugs. As an indicator of liver and kidney disease, immune deficiencies, malignancies of the immune system, acute and chronic infection, genetic deficiencies, central nervous system disease, and numerous other pathologies, serum protein electrophoresis exceeds any other single test in its ability to detect so broad a spectrum of disease states.

# 1.2   Materials

## 1.2.1   Reagents and solutions

(1) Sample: fresh serum.

(2) Barbiturate buffer (1 L, pH 8.6) : 1.66 g barbital, 12.76 g barbital sodium, qs to 1 L with Milli-Q $H_2O$.

(3) Amino black 10B staining solution: 0.5 g amino black 10B, 50 mL methanol, 10 mL glacial acetic acid, qs to 100 mL with Milli-Q $H_2O$.

(4) Destaining solutions: 45 mL alcohol, 5 mL glacial acetic acid, qs to 100 mL with Milli-Q $H_2O$.

(5) Leaching agent: 1.6 g sodium hydroxide, qs to 100 mL with Milli-Q $H_2O$.

## 1.2.2   Special equipment

①Electrophoresis chamber; ②power supply; ③micropipettes; ④spectrophotometer; ⑤cellulose-acetate membrane.

# 1.3   Methods

(1) Prepare cellulose-acetate membrane

1) Preparea 2 cm×8 cm cellulose acetate membrane, draw a line gently with pencil on the smooth side of the membrane, about 1.5 cm far away from one short edge.

2) Saturate the cellulose-acetate membrane by immersing in *Barbiturate* buffer for 10-20 min with smooth surface upward. Slightly dry the strip on a filter paper sandwich.

(2) Load the sample: load 2-3 μL serum on the back of the pencil line on the rough surface using a sampler (this is the most important step in obtaining good results). Wait for a moment and let serum be absorbed by the membrane. Put the strip downward across the two tanks of the chamber, with the loading end near the cathode, using filter papers as wicks. Balance the strip for 5 min.

(3) Electrophorese the sample: set current on 80-100 v[ (10 v/m · strip) ], 0.4-0.6 mA/(m · strip). Electrophoresis is carried on for 40-60 min.

(4) Stain and destain

1) Then switch off the power and remove strips from the chamber.

2) Stain the strip by immersing in staining solution for 10 min.

3) Stepwise rinse in destaining solutions for 10 min twice, the sandwich strip with filter paper carefully to drain solution.

4) Observe the shade of width and order of the bands and record them.

(5) Separate protein bands

1) Dissect the five major identified serum protein bands with scissors carefully.

2) Individual protein band and one blank band are all put in tubes with 4 mL leaching agent, respectively. Warm them at 37 ℃ for 30 min with un-continuous shaking.

(6) Measure the absorbance

1) Read the absorbance of each protein at 650 nm against the blank on a spectrophotometer.

2) Record the absorbance: $A_a$ of albumin, $A_{\alpha1}$ of $\alpha_1$-globulin, $A_{\alpha2}$ of $\alpha_2$-globulin, $A_\beta$ of $\beta$-globulin, and $A_\gamma$ of $\gamma$-globulin.

# 1.4   Results and discussion

(1) Result of CAME

Electrophoresis results (Figure 8.1, Figure 8.2).

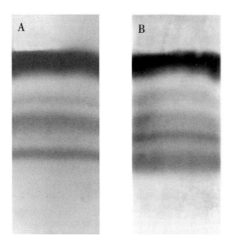

Figure 8.1   **CAME good images**

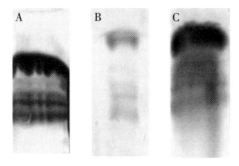

Figure 8.2   **CAME bad images**

(2) Calculation of ratios

Absorbance summation $(S)$: $S = A_a + A_{\alpha1} + A_{\alpha2} + A_\beta + A_\gamma$

The ratio of albumin $= A_a/S \times 100\%$

The ratio of $\alpha_1$-globulin $= A_{\alpha1}/S \times 100\%$

The ratio of $\alpha_2$-globulin $= A_{\alpha2}/S \times 100\%$

The ratio of $\beta$-globulin $= A_\beta/S \times 100\%$

The ratio of $\gamma$-globulin $= A_\gamma/S \times 100\%$

(3) Reference value

The ratio of albumin: 57.45%-71.73%.

The ratio of $\alpha_1$-globulin: 1.76%-4.48%.

The ratio of $\alpha_2$-globulin: 4.04%-8.28%.

The ratio of $\beta$-globulin: 6.79%-11.39%.

The ratio of $\gamma$-globulin: $11.85\% - 22.97\%$.

## 1.5    Troubleshooting

Problems encountered in CAME and guidance for how to address major difficulties (Table 8.1).

Table 8.1    CAME troubleshooting

| Problem | Potential causes | Solutions |
|---|---|---|
| Smeared protein gels | Gross overloading<br>Overloaded, nonuniform gel<br>Failure to denature sample | · Do not gross overloading of protein<br>· Samples had to be diluted prior to preparing a protein assay tube |
| Streaked protein gels | Too wet CAM<br>Overloading | · Proper humidity<br>· Do not gross overloading of protein |
| Bands too weak | Sloppy loading<br>Bad staining<br>Forgot to stain | · The sample should be properly drawn up into the syringe<br>· By re-staining with fresh dye |
| Bands only at top of the gel | Stopped early, diffuse dye front<br>Stopped early, tight dye front | · Improved with the distance the proteins are permitted tomigrate until resolution becomes limited by diffusion<br>· The current is 0.4-0.6 mA/cm |

# Experiment 2    Separation and Determination of Protein by SDS-polyacrylamide Gel Electrophoresis

## 2.1    Principle

The electrophoretic separation of proteins is usually accomplished using polyacrylamide gel electrophoresis (PAGE), in which the proteins are driven by an applied current through agelated matrix. The matrix is composed of polymers of a small organic molecule (acrylamide and bisacrylamide) that is cross-linked to form a molecular sieve. The pore size is determined by the ratio of acrylamide to bisacrylamide, and the concentration of acrylamide. Polymerization of acrylamide and bisacrylamide monomers is induced by ammonium persulfate (AP), which spontaneously decomposes to form free radicals. TEMED, a free radical stabilizer, is generally included to promote polymerization.

PAGE is usually carried out in the presence of the negatively charged detergent sodium dodecyl sulfate (SDS), which binds in large numbers to all types of protein molecules. The electrostatic repulsion between the bound SDS molecules causes the proteins to unfold into a similar rod-like shape, thus eliminating differences in shape as a factor in separation. In the presence of SDS, the intrinsic charge of a protein is masked. As a result, proteins are separated by SDS-PAGE on the basis of a single property—their molecular mass.

# 2.2　Materials

## 2.2.1　Reagents and solutions

(1) The sample solution: the recombinant clone *E. coil* BL-21.

(2) 2×Sample Buffer( 10 mL) : 0. 15 g tris base, 1. 0 mL β-Mercaptoethanol, 0. 01 g bromophenol blue, 2. 0 mL glycerol, 0. 4 g SDS, 7. 0 mL distilled water.

Divide into 0. 5-1. 0 mL aliquots and store at -80 ℃. Working solution can be stored at -20 ℃. Add an equal volume of sample buffer to the sample and heat at 100 ℃ in a heat block for 4 min.

(3) Prestained molecular weight marker.

1) 1×TE Buffer (500 mL) : 5 mL of 1 mol/L Tris-HCl/pH 8. 0, 1 mL of 0. 5 mol/L EDTA/pH 8. 0, qs to 500 mL with distilled water.

2) 30% acryl-bisacrylamide ( acry : bis 29 : 1 ) : 29 g acrylamide, 1 g bisacrylamide, qs to 100 mL with distilled water. Store in the dark at 4 ℃ ( wrap bottle in foil) for 3 months. It is best to be used in one month. Acrylamide and bisacrylamide powders are neurotoxins.

3) 10% ammonium persulfate ( AP) : 0. 1 g AP in 1 mL distilled water, prepare a 10% AP solution fresh daily in a microfuge tube and add the appropriate amount of water.

(4) TEMED: Store in the dark at RT.

1) 1×tris-glycine running buffer ( 1 L) : 25 mmol/L Tris 3. 02 g, 250 mmol/L glycine 18. 8 g, 0. 1% SDS 1 g, qs to 1 L with distilled water.

2) Coomassie blue staining solution( 100 mL) : 0. 1% Coomassie blue in 45 : 45 : 10 methanol : water : acetic acid.

3) Destaining solution( 100 mL) : 45 : 45 : 10 methanol : water : acetic acid.

## 2.2.2　Special equipment

①Mini-Protein electrophoresis equipment; ②power supply; ③micropipettes; ④microfuge tubes.

# 2.3　Methods

## 2.3.1　Assemble the glass plate

(1) Clean and dry the glass plates.

(2) Select a proper spacer plate of the desired gel thickness and place a short plate on top of it ( Figure 8. 3a) .

(3) Orient the spacer plate so that the labeling is up. Slide the two glass plates into the casting frame, keeping the short plate facing the front of the frame ( Figure 8. 3b) .

(4) When the glass plates are in place, engage the pressure cams to secure the glass cassette sandwich in the casting frame ( Figure 8. 3c) . Check that both plates are flush at the bottom.

(5) Place the casting frame into the casting stand by positioning the casting frame onto the casting gasket while engaging the spring-loaded lever of the casting stand onto the spacer plate ( Figure 8. 3d) .

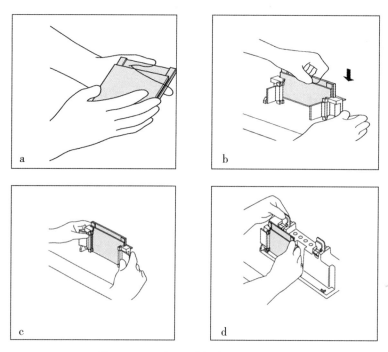

Figure 8.3 **Assemble the glass plate**

## 2.3.2 Pour the gels

(1) Prepare 10% separating/resolving gel and 5% stacking gel as directed in Table 8.2.

Table 8.2 **Preparation of SDS-PAGE gel**

| 10% Separating/resolving gel (10 mL) | | 5% Stacking gel (5 mL) | |
| --- | --- | --- | --- |
| $H_2O$ | 4.0 mL | $H_2O$ | 3.4 mL |
| 30% acryl-bisacrylamide mix | 3.3 mL | 30% acryl-bisacrylamide mix | 0.83 mL |
| 1.5 mol/L Tris (pH 8.8) | 2.5 mL | 1.0 mol/L Tris (pH 6.8) | 0.63 mL |
| 10% SDS | 0.1 mL | 10% SDS | 0.05 mL |
| 10% ammonium persulfate | 0.1 mL | 10% ammonium persulfate | 0.05 mL |
| TEMED | 0.004 mL | TEMED | 0.005 mL |

(2) Prepare the separating gel monomer solution by combining all reagents except AP and TEMED. Deaerate and mix the solution after adding each reagent by swirling the container gently. Place a comb completely into the assembled gel cassette. Mark the glass plate 1 cm below the comb teeth.

(3) Add AP and TEMED to the degassed monomer solution and pour to the mark.

(4) Immediately overlay the monomer solution with water or t-amyl alcohol.

(5) Allow the gel to polymerize for 45 min to 1 h. Rinse the gel surface completely with distilled water. Do not leave the alcohol overlay on the gel for more than 1 h because it will dehydrate the top of the gel.

(6) Prepare the stacking gel monomer solution. Combine all reagents except AP and TEMED. Deaerate and mix the solution after adding each reagent by swirling the container gently.

(7) Add AP and TEMED to the degassed stacking gel monomer solution and pour the solution between the glass plates. Continue to pour until the top of the short plate is reached.

(8) Place a comb in the gel sandwich. Allow the gel to polymerize for 15 min. Remove the comb.

(9) The gel is placed in the buffer chamber and running gel buffer is added into the chamber.

## 2.3.3　Prepare the samples

From the recombinant clone *E. coil* BL-21, the recombinant protein is produced as follows. The clone had grown for 4 hours and was induced using IPTG for 5 hours at 30 ℃. The culture was pelleted and resuspended in 1×TE buffer. Each group is provided as Table 8.3.

Table 8.3　Preparation of samples

| Procedure | Tube 1 | Tube 2 | Tube 3 |
| --- | --- | --- | --- |
| Content | Mol. Wt Marker | Uninduced clone protein | Induced clone protein |
| 2×Sample buffer | – | 20 | 20 |
| Boil | 5 min | 5 min | 5 min |
| Brief spin | Not required | 5 min | 5 min |
| Loading sample into the well | 10 | 20 | 20 |

## 2.3.4　Run the gel

(1) Load equal amounts of protein into the wells of the SDS-PAGE gel, along with molecular weight marker.

(2) Attach the electrical leads to a suitable power pack with the proper polarity (black to black and red to red), and run the gel until the dye front of bromophenol blue has migrated to the very bottom of the gel.

## 2.3.5　Disassemble the gel and stain

(1) After the electrophoresis is complete, turn off the power supply and disconnect the electrical leads. Remove the tank lid and carefully lift out the electrode assemblies. Pour off and discard the running buffer.

(2) Open the arms of the assembly and remove the gels from the gel cassette by gently separating the two plates of the gel cassette.

(3) Place the gels in the container containing the Coomassie blue staining solution. Make sure that the gel is fully submerged in the staining solution.

(4) Stain the gel for 1 h, agitate it slowly on a shaker.

## 2.3.6　Destain the gel and analyze

(1) Transfer the gel (save the dye mixture; it can be re-used many times) to the destaining solution, place on a shaker, and replace with fresh rinse mixture until the excess dye has been removed.

(2) Put your destained gel on a piece of saran wrap or in Ziploc bag and photograph it with a digital camera. Include a centimeter ruler in your photograph so that you can easily quantify your measurements. You should put a print of this image into your notebook.

(3) Use an image modifying the software to make a grayscale TIF file out of your image and save it as "SDS-gel. TIF". Save the file somewhere on your computer where you can easily find it when needed.

(4) Use Scion Image(NIH Image) to measure the migration distances on of the various proteins on your gel. You need to analyze the grayscale "SDS-gel. TIF" file and to set the scale in millimeters using the ruler that you included in your photograph. Once the scale is set, start your measurements at the beginning of the resolving gel and end them at the forward edge of the band under consideration. Remember to measure the migration distance of the bromophenol blue for each lane so that you can calculate the $R_f$ for each band. Record all of your data in a spreadsheet for easy data manipulation and graphing. You may print a copy of your

spreadsheet and put it into your notebook.

## 2.4 Results and discussion

(1) Result of SDS—PAGE (Figure 8.4).

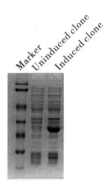

Figure 8.4  **SDS—PAGE gel image**

(2) Determine the molecular weights of the proteins in the sample.

To determine the molecular weights of the denatured polypeptides you will plot a standard curve based on the linear relationship between the log of molecular weight of the protein standards and the $R_f$ for each standard polypeptide as detailed below.

For each protein in the SDS—PAGE molecular weight standards, you have already used Scion Image to measure the distance in millimeters from the top of the resolving gel to the top of each band. You have also measured the distance from the top of the resolving gel to the top of the bromophenol blue ( dye front) band and recorded the measurements in a spreadsheet.

1) Calculate the $R_f$ ( ratio of the fronts) of each protein standard, using the equation: $R_f$ = distance of protein migration/distance of dye front migration.

2) Using your computer, plot the log of the molecular weight on the axis and the $R_f$ on the $x$—axis. Have your computer draw the best fit line through the data points. Print your standard curve and put it in your notebook. Also, be able to graph your results by hand using two—cycle semi—log paper. This is important because you might be called on to do a similar exercise as part of the laboratory exam. If you wish to, you may make up the same standard curve on semi—log paper and put it your notebook too.

3) Determine the molecular weight of several other polypeptides run on the same gel by measuring the migration distance of the band, calculating the $R_f$, and then use the standard curve to read the log of molecular weight directly from the graph.

(3) What further test( s) would you do to establish that a particular protein band in one fraction is identical to ( or different from) a particular band in another fraction.

## 2.5 Troubleshooting

Problems encountered in SDS—PAGE and guidance for how to address major difficulties ( Table 8.4).

Table 8.4   SDS-PAGE troubleshooting

| Problem | Possible cause | Solution |
|---|---|---|
| Smeared protein gels | Bad pour<br>Gross overloading<br>Overloaded, nonuniform gel<br>Failure to denature sample | · Evenly poured acrylamide mixture<br>· Do not gross overloading of protein<br>· Samples had to be diluted prior to preparing a protein assay tube |
| Streaked protein gels | Particles in sample<br>Overloading<br>Uneven top of the gel | · The electric field must be quite uniform<br>· Must use an overlay after pouring the first gel layer |
| Bands too light | Sloppy loading<br>Bad stain<br>Overestimated protein concentration<br>Forgot to stain | · Be precipitated in the gel by the acidified alcohol<br>· By re-staining with fresh dye<br>· The sample must be prepared to the correct concentration for electrophoresis<br>· The sample should be properly drawn up into the syringe |
| Bands only at top of the gel | Stopped early, diffuse dye front<br>Stopped early, tight dye front | · Improved with the distance the proteins are permitted to migrate until resolution becomes limited by diffusion |
| Miscellaneous | Crooked bands<br>Horizontal lines<br>No penetration<br>Broken gel<br>Field effects | · The surface of the resolving gel must be even<br>· The gel mix should be clearly correct<br>· Keep in mind that the gel will still be usable as long as you save the pieces<br>· Make sure the electrodes are inserted correctly |
| Multiple symptoms | No dye front, crooked, overloaded field effects<br>Torn gel, uneven loading | · Electrophoresis could not be terminated too late<br>· Even resolving gel surface<br>· Even loading |
| Bizarre results | Wrong electrode buffer<br>Overheating<br>Delay in running gels<br>Moldy buffer<br>Opened gel box during the run | · Prepare the correct electrophoretic buffer<br>· After sampling is finished, the electrophoresis should be carried out in time<br>· Electrophoresis tank can be placed in an ice water mixture to cool the electrophoretic buffer |

# Experiment 3   Identification of Target Protein by Western Blotting

## 3.1   Principle

Western blot is also called immunoblotting. It is a kind of immunochemical techniques which is used to detect a protein immobilized on a matrix. The target protein can be in a crude extract or a more purified preparation and the monoclonal or polyclonal antibody against this protein is necessary to help us to recognize the antigen. Western blot could detect target protein which is as low as 1 ng due to the high resolution of the gel electrophoresis and strong specificity and high sensitivity of the immunoassay.

In this method, proteins are separated by SDS−PAGE and transferred from the gel to a hydrophobic or positively charged membrane. The membrane is then incubated with specific antibodies in the presence of a protein−rich blocking agent such as a solution containing BSA, which reduces non−specific binding of the antibody. After the membrane is washed, the bound antibody can be detected in a variety of ways, most of which rely on "secondary" antibodies that bind to the "primary" antibody. These "secondary" antibodies themselves can be directly labeled with radioactivity, coupled to a fluorescent molecule, or linked to an enzyme such as horseradish peroxidase, which produces a colored or fluorescent band at the location of the protein of interest when incubated with an appropriate substrate.

Generally, two classes of "primary" antibodies are used: polyclonal antibodies, which are a collection of different antibodies that often react with different epitopes of the same protein, and monoclonal antibodies, which are obtained from a clonal population of immune cells that express only one type of antibody and thus react with one epitope of the protein.

# 3.2 Materials

## 3.2.1 Reagents and solutions

(1) 1×transfer buffer(1 L): 25 mmol/L Tris 3.02 g, 250 mmol/L glycine 18.8 g, Methanol 200 mL, qs to 1 L with distilled water.

(2) 1×PBST buffer (1 L): 135 mmol/L NaCl 7.9 g, 2.7 mmol/L KCl 0.2 g, 10 mmol/L $Na_2HPO_4$ 1.42 g, 10.3 mmol/L $K_2HPO_4$ 1.8 g, check the pH and adjust to 8.3, qs to 1 L with distilled water.

(3) Blocking buffer: 3%−5% milk or BSA (bovine serum albumin), add to PBST buffer. Mix well and filter.

(4) ECL Western Blotting Substrate: from Thermo Fisher Scientific Inc.

## 3.2.2 Special equipment

①Mini−Proteintransfer equipment; ②power supply; ③micropipettes; ④filter paper; ⑤PVDF membranes; ⑥forceps; ⑦chemiluminescence analyzer.

# 3.3 Methods

## 3.3.1 Prepare blotting

(1) After the electrophoresis is complete, turn off the power supply. Remove the gels from the gel cassette by gently separating the two plates of the gel cassette. Prepare the transfer buffer(using buffer chilled to 4 ℃ will improve heat dissipation).

(2) Cut the membrane and the filter paper to the dimensions of the gel or use precut membranes and filter paper. Always wear gloves when handling membranes to prevent contamination. Equilibrate the gel and soak the membrane, filter paper, and fiber pads in transfer buffer (15−20 min depending on gel thickness).

(3) Prepare the gel sandwich. Place the cassette, with the gray side down, on a clean surface.

(4) Place one prewetted fiber pad on the gray side of the cassette.

(5) Place a sheet of filter paper on the fiber pad.

(6) Place the equilibrated gel on the filter paper.

(7) Place the prewetted membrane on the gel.

(8) Complete the sandwich by placing a piece of filter paper on the membrane.

(9) Add the last fiber pad.

(10) Close the cassette firmly and be careful not to move the gel and filter paper sandwich. Lock the cassette closed with the white latch.

(11) Place the cassette in the module. Repeat for the other cassette.

(12) Place in the tank and fill with pre-cooling transfer buffer.

(13) Put on the lid, plug the cables into the power supply, and run the blot.

### 3.3.2 Block

(1) Upon completion of the running, disassemble the blotting sandwich and remove the membrane for development. Clean the cell, fiber pads, and cassettes with laboratory detergent and rinse well with deionized water.

(2) Block the membrane for 1 h at room temperature (RT) using blocking buffer.

### 3.3.3 Immunodetection

(1) Cut the membrane into strips if necessary and prepare the primary antibody by diluting in blocking buffer at the appropriate dilution.

(2) Incubate the membrane with the primary antibody for 2 hours at RT or overnight at 4 ℃ (recommended).

(3) Rinse the blots briefly in PBST and then perform three times washes for 10 min using the same buffer.

(4) Incubate the membrane with the secondary antibody for 1 h at RT.

(5) Wash the membrane in PBST three times for 10 min.

(6) Perform ECL as described by the manufacturer, e. g., add ECL reagents for 3 min at RT and capture WB image using various durations of exposure: 10 s, 30 s, 1 min, 2 min, 3 min, 4 min, and 5 min.

## 3.4 Results and discussion

(1) The result of western blot (Figure 8.5).

Figure 8.5 **Western blot image**

(2) Analysis: design of proper control is essential to Western blot. It will guarantee an accurate and specific result by identifying various problems quickly and precisely.

(3) Control types

1) Positive control: a lysate from a cell line or tissue sample known to express the protein you are detecting. The positive control is designed to verify the working efficiency of the antibodies.

2) Negative control: A lysate from a cell line or tissue sample known not to express the protein you are detecting. Negative control is to check antibody specificity. Nonspecific binding and a false positive result will be identified.

3) Secondary antibody control (no primary antibody control): the primary antibody is not added to the membrane. The only secondary antibody is added. This is to check secondary antibody specificity. Nonspe-

cific binding and the false positive result caused by secondary antibody will be indicated.

4) Blank control: both primary and secondary antibody are not added to the membrane. This is to check membrane nature and blocking effect.

5) Loading control: loading control is used to check sample quality and the performance of secondary antibody system.

(4) Loading controls are antibodies to "house–keeping proteins", or proteins that are expressed at equivalent levels in almost all tissues and cells. Loading controls are required to check that the lanes in your gel have been evenly loaded with the sample, especially when a comparison must be made between the expression levels of a protein in different samples. They are also useful to check for even transfer from the gel to the membrane across the whole gel. Where even loading or transfer have not occurred, the loading control bands can be used to quantify the protein amounts in each lane. For publication–quality level, the use of a loading control is absolutely essential.

# 3.5   Troubleshooting

Problems encountered in Western blot and guidance for how to address major difficulties (Table 8.5).

Table 8.5   Western blot troubleshooting

| Problem | Possible cause | Solution |
| --- | --- | --- |
| No signal | The primary and secondary antibody is not compatible | · Use acorrect secondary antibody that was raised against the species in which the primary was raised |
| | There is an insufficient antigen | · Load at least 20–30 μg protein per lane, use protease inhibitors and run the recommended positive control |
| High background | Blocking of non–specific binding might be absent or insufficient | · Increase the blocking incubation period and consider changing the blocking agent |
| | The primary antibody concentration may be too high | · Titrate the antibody to the optimal concentration. Incubate for longer but in more dilute antibody |
| | The wash of unbound antibodies may be insufficient | · Increase the number and time of washes |
| Multiple bands | Primary antibody concentration is too high | · Try decreasing the concentration of the primary antibody. Run a secondary antibody control (without the primary) |
| | The bands may be non–specific | · Where possible use blocking peptides to differentiate between specific and nonspecific bands |
| Uneven white spots in the blot | Air bubbles were trapped against the membrane during transfer or the antibody is not evenly spread on the membrane | · Make sure you remove bubbles when preparing the gel for transfer. Incubate the antibodies while agitating |
| Smile effect on the bands | Migration was too fast | · Decrease the voltage while running the gel |
| | Migration was too hot | · Run the gel in the cold room or on ice |

Continue to Table 8.5

| Problem | Possible cause | Solution |
| --- | --- | --- |
| Band of interest is very low/high on the blot | Separation is not efficient | · Change the gel percentage: use a higher percentage for small proteins and a lower the percentage for large proteins |
| Uneven band size in lanes probed for the same protein | Gel has set too quickly while casting and the acrylamide percentage is not even throughout the gel | · Review the recipe of the gel and the addition of TEMED to the gels, add some 0.1% SDS in water to the top of the migrating gel while it set to stop it from drying |

*Du Jian, Fan Xinjiong*

# Chapter 9

## Separation and Identification of Proteins and Amino Acids by Chromatography

### Experiment 1   Thin Layer Chromatography (TLC) of Amino Acids

## 1.1   Principle

Thin layer chromatography is one of the chromatography methods used in separating mixtures into different components. Similar to all the other chromatography, it also has a stationary phase and a mobile phase. Normally, in TLC, the stationary phase is a layer of silica gel coated on a sheet of glass, while the mobile phase can be a single solvent or solvent mixture. Here in this experiment, we will use the mixture of methyl alcohol and glacial acetic acid as the solvent.

The silica gel has a different absorption capacity for different amino acids, the interaction between eluent and different amino acids is also different. Therefore, different amino acids can be separated according to their different migration. To quantify the results, the retardation factor ($R_f$) is usually explored to express the migration ability for different component (see Chapter 5).

The separated amino acids are visualized using a ninhydrin solution. Amino acids contain a free amino and carboxyl group which reacts together with ninhydrin to produce a characteristic blue color (or occasionally pale yellow).

## 1.2   Materials

### 1.2.1   Reagents and solutions

(1) Mobile phase(solvent): a mixture of 80% butanol, 10% acetic acid, and 10% water.

(2) Amino acids solution: solutions of the amino acids.

1) Alanine (0.01%): 8.9 mg of alanine is dissolved in 10 mL of isopropanol (90%).

2) Arginine (0.01%): 17.4 mg of arginine is dissolved in 10 mL of isopropanol (90%).

3) Glycine (0.01%): 7.5 mg of glycine is dissolved in 10 mL of isopropanol (90%).

(3) Test sample: the mixtures of the two amino acids are mixed in equal volume.

(4) Neutral triketohydrindene solution (0.2%): 0.2 g of triketohydrindene hydrate dissolved in 100 mL of acetone (developing reagent).

## 1.2.2  Special equipment

①TLC plates with G-silica gel (6 cm, 15 cm); ②beaker (50 mL); ③capillary tube; ④ruler; ⑤hair drier.

# 1.3   Methods

## 1.3.1   Flow chart

The TLC flow chart B shown in Figure 9.1.

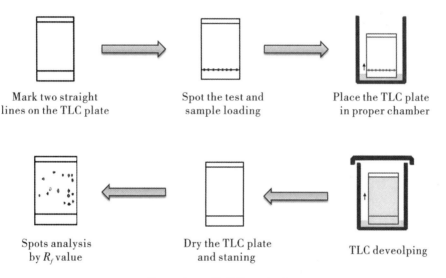

<table>
<tr><td>Mark two straight<br>lines on the TLC plate</td><td>Spot the test and<br>sample loading</td><td>Place the TLC plate<br>in proper chamber</td></tr>
<tr><td>Spots analysis<br>by $R_f$ value</td><td>Dry the TLC plate<br>and staning</td><td>TLC deveolping</td></tr>
</table>

Figure 9.1   **TLC flow chart**

## 1.3.2   Procedure

(1) Take a TLC plate and with the help of pencil draw two straight lines on the white surface of the plate: one 2 cm from the bottom of the plate and another 1 cm from the top of the plate.

(2) Mark several equidistant points on the bottom line for loading of amino acids samples and test sample. While marking the lines and points do not make a trough witha pencil.

(3) Allow all the amino acid samples and test sample to come to room temperature. Using a capillary tube to spot each amino acid and the test sample along the bottom line on the TLC plate. While spotting use separate tips for each sample.

(4) Allow the plates for air drying (−10 min). Further drying should be done by keeping the TLC plate at 70 ℃ in a hot air oven or incubator for 2−3 min.

(5) Take 10 mL solvent system in the TLC chamber (with lid) and wait for 10 min at room temperature.

(6) Place the TLC plate inside the chamber with a clean forceps. Make sure that the spotted samples are near the solvent. Furthermore, the TLC plate should be in a straight position so that the solvent phase

can move uniformly along the plate.

(7) Allow the solvent front to reach the top line of the plate. After that take it out with the help of clean forceps and air dry the plate for 15–20 min. Keep the plates at 70 ℃ for 2 min for further drying.

(8) Add 1 mL the developing reagent on the plate and swirl the plate very carefully. Look for the development of the colored spots of different amino acids and the test sample.

(9) After the plate is air dried, $R_f$ values are calculated for each amino acid and different amino acids in the test sample.

## 1.4　Results and discussion

(1) Staining results: after drying the TLC plate in the oven, five blue–purple spots appear on the plate, and the results can be seen in Figure 9.2.

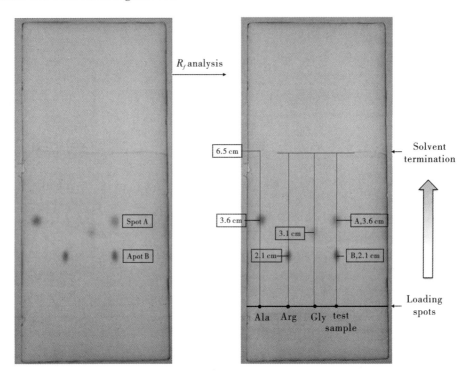

Figure 9.2　$R_f$ analysis and determining the test sample

(2) Original data: the migration distance for each spot as well as the solvent can be measured, and listed in Table 9.1.

Table 9.1　Original data of the certain migration distances

| Loading spot | Sample migration distance | Solvent migration distance |
| --- | --- | --- |
| Alanine | 3.6 cm | 6.5 cm |
| Arginine | 2.1 cm | 6.5 cm |
| Glycine | 3.1 cm | 6.5 cm |
| Spot A | 2.1 cm | 6.5 cm |
| Spot B | 3.6 cm | 6.5 cm |

(3) Determination of the test sample: based on the migration distance data, $R_f$ values of each amino acids and spots in the test sample can be calculated. The amino acids in the test sample can be determined according to certain $R_f$ values (Table 9.2).

Table 9.2 Determination of the test sample

| Loading spot | $R_f$ value | Amino acids |
| --- | --- | --- |
| Alanine | 0.55 | — |
| Arginine | 0.32 | — |
| Glycine | 0.48 | — |
| Spot A | 0.32 | Arginine |
| Spot B | 0.55 | Alanine |

(4) Result: within the certain range of error, $R_f$ value analysis indicates that the amino acids in the test sample solution are arginine and alanine.

## 1.5 Troubleshooting

Problems encountered in thin layer chomatography of amino acicls and guidcnce for how to address major difficulties (Table 9.3).

Table 9.3 TCL Troubleshooting

| Problem | Possible cause | Solution |
| --- | --- | --- |
| The appearance of over-large spots after development | The initial spot is larger than 2 mm in diameter | While spotting the samples on the TLC plate ensure that the spots are not larger than 1–2 mm in diameter |
| Solvent front advances unevenly | Use of a developing chamber that does not have a flat bottom | It is therefore important to use flat bottomed developing tanks during TLC |
| The substance moves along the TLC plate as a long streak, rather than as a single discrete spot | Spotting the plate with too much of sample, more than the moving solvent can handle | Streaking can be eliminated by systematically diluting the spotting solution until development and visualization show the substances moving as single spots, rather than elongated streaks |
| Irregular staining spots on TLC plate | Mixture solution left on hands might absorb onto the plate | Never touch the surface of the plate by hands |

# Experiment 2 Paper Chromatography of Carotene

## 2.1 Principle

Paper chromatography is a technique used to separate substances in a mixture based on the movement of the different substances up to a piece of paper by capillary action. The hydroxyl on the chromatography paper can absorb a layer of water and become the stationary phase, while the mobile phase is generally an alcohol solvent mixture. The process of paper chromatography is similar to TLC. The retardation factor ($R_f$) is also used to identify different components in the test sample according to the known $R_f$ value of the standard sample. Details for $R_f$ value see Chapter 9 Experiment 1.

Plant pigments are macromolecules produced by the plant, and these pigments absorb specified wavelengths of visible light to provide the energy required for photosynthesis. A small sample of plant pigment placed on chromatography paper travels up the paper due to capillary action. Beta carotene is carried the furthest because it is highly soluble in the solvent and because it forms no hydrogen bonds with the chromatography paper fibers. Xanthophyll contains oxygen and does not travel as far with the solvent because it is less soluble than beta carotene and forms some hydrogen bonds with the paper. Chlorophylls are bound more tightly to the paper than the other two, so they travel the shortest distance.

## 2.2   Materials

### 2.2.1   Reagents and solutions

①Acetone; ②70% Isopropyl alcohol; ③chromatography solvent (9 : 1 petroleum ether & acetone).

### 2.2.2   Special equipment

①2 or 3 fresh spinach leaves; ②ruler; ③large test tube; ④cork with push pin; ⑤chromatography paper (precut 18 cm strips); ⑥pencil and colored pencils; ⑦coin; ⑧5 mL syringe; ⑨calculator; ⑩scissors; ⑪plastic wrap; ⑫plastic pipettes.

## 2.3   Methods

### 2.3.1   Safety

(1) Wear goggles and aprons when working with chemicals.
(2) Petroleum ether, acetone, and alcohol are volatile and flammable.
(3) Avoid breathing vapors of the reagents.

### 2.3.2   Procedure

(1) Prepare the Sample
CAUTION! Oil from the skin affects the separation, so handle paper as little as possible and only by the edges.

1) Take a strip of chromatography paper approximately 18 cm long. One end is blunt and the other is pointed.

2) With a pencil lightly make a line 2 cm from the pointed end of the paper( Figure 9.3).

Figure 9.3   Prepare the sample A

3) Bend the strip of paper at the blunt end and attach it to the small end of the cork with the push pin. Adjust the length of the paper so that when it is inserted into the test tube, it will touch the bottom without curling( Figure 9. 4) .

Figure 9. 4    **Prepare the sample B**

4) Place a ruler over the leaf so that covers the pencil line on either end( Figure 9. 5) .

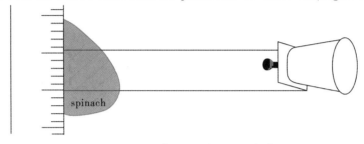

Figure 9. 5    **Prepare the sample C**

5) Using a coin, press down firmly and roll along the ruler edge several times to form a definite green line.

6) Allow the green line to dry thoroughly.

7) Use a fresh area of the leaf and repeat several times until the pencil line is covered completely with a narrow green band. Be careful not to smear this green line.

( 2) Separate the pigments

1) Place the test tube in the test tube rack. Using the 5 mL syringe, dispense 5 mL of chromatography solvent in the test tube.

2) Carefully lower the paper strip into the test tube and secure the cork in the top. The solvent must touch the pointed end of the paper but should not touch the green line.

3) Be careful not to slosh the solvent. Allow the tube to stand undisturbed.

4) Observe the solvent movement and band separation.

5) When the pigments have separated into distinct bands ( the solvent has moved approximately half the distance of the paper) , lift the cork with paper attached from the test tube. Mark the edge of the solvent front with a pencil. Remove the push pin and detach the paper from the cork.

6) Place the push pin back in the cork and place the cork back on the test tube to minimize fumes.

7) Allow the paper to dry completely.

## 2. 4    Results and discussion

( 1) On the report book, color the diagram to illustrate the colored bands on the chromatogram. Label the band that traveled the greatest distance 1, the next 2, the next 3, and continue until all bands are labeled ( Figure 9. 6) .

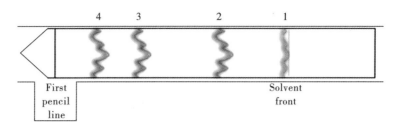

Figure 9.6    Label the bands

(2) Describe the color of each band in Table 9.3(column B).

    (3) Measure the distance from the first pencil line to the solvent front. Record this value in Table 9.3 (column C) for each pigment.

    (4) Measure the distance from the first pencil line to the average peak of each color band( Figure 9.7), and record these values in Table 9.3 (column D).

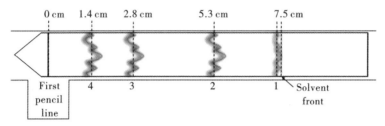

Figure 9.7    Measurement of the distance

    (5) Calculate the $R_f$ values for each pigment and record the values in Table 9.3 (column E) using the $R_f$ calculation formula.

Table 9.3    Results, calculation, and determination of each component.

| Entry | Color | Migration | $R_f$ value | Component |
|---|---|---|---|---|
| Solvent front | – | 7.5 cm | | |
| 1 | | | | |
| 2 | | | | |
| 3 | | | | |
| 4 | | | | |

Note: $R_f$ Values for known substance: β−carotene, 0.99; chlorophy Ⅱ a, 0.30; chlorophy Ⅱ b, 0.13; violaxanthin, 0.40; lutein, 0.68. Carotene, violaxanthin, and lutein are carotenoids.

## 2.5　Troubleshooting

    Problems ercountered in paper chromatography of carotene and guidance for how to address major difficulties (Table 9.4).

Table 9.4    Troubleshooting of paper chromatography of carotene

| Problem | Possible cause | Solution |
|---|---|---|
| The appearance of overwide bands after development | Initial loading band of pigments is too wide | While loading the spinach pigments on the paper ensure that the green line is no wider than 1–2 mm |
| Only one colored band is found beside the loading line | The eluent polar is inappropriate | More acetone is needed in the eluent solution |
| Pigments move along the paper as a sheet rather than as a single colored band | The pigments are over loaded. Or, the loading line is immersed by the eluent solvent | Avoid loading too many pigments with the spinach leaf <br> Make the loading line thoroughly dried before putting the paper in the test tube <br> Make sure not to allow the loading line getting into the solvent |

# Experiment 3    Ion–exchange Chromatography of Mixed Amino Acid

# 3.1    Principle

Ion–exchange resin is a kind of synthetic polymer. Related principle please see Chapter 5. It's relatively ideal to use ion–exchange resin to separate small molecules such as amino acid, adenosine, and nucleotide. But it's not proper for a macro–molecule substance like protein, because it can't diffuse into the chain structure of resin. So ion–exchange reagent such as cellulose and dextran as the supporter can be chosen to separate biological macromolecules.

This experiment uses sulfonic acid cationic exchange resin to separate the mixture of acidic amino acid (aspartic acid), neutral amino acid (alanine) and basic amino acid (lysine). In the condition of a certain pH, the degree of their dissociation is different, so they can be eluted and separated by changing the pH and the ionic strength of the eluting solution, respectively.

# 3.2    Materials

## 3.2.1    Reagents and solutions

(1) Hydrogen chloride (2 mol/L).

(2) Sodium hydroxide (2 mol/L).

(3) Hydrogen chloride (0.1 mol/L).

(4) Sodium hydroxide (0.1 mol/L).

(5) Citric acid buffer (pH 4.2): 54 mL of 0.1 mol/L citric acid and 46 mL of 0.1 mol/L sodium citrate.

(6) Acetic acid buffer (pH 5): 70 mL of 0.2 mol/L sodium acetate and 30 mL 0.2 mol/L acetate acid.

(7) Neutral triketohydrindene hydrate solution (0.2%): 0.2 g of triketohydrindene hydrate and 100 mL of acetone.

(8) Mixture of amino acids: 10 mL of each of alanine, aspartic acid, lysine and 3 mL of 0.1 mol/L hy-

drogen chloride.

## 3.2.2 Special equipment

①Column(16/20); ②pump; ③gradient maker; ④ultraviolet meters; ⑤sulfonic acid cationic exchange-resin.

# 3.3 Methods

(1) Preparing the resin

Put about 10 g of resin in a 100 mL beaker, add 25 mL of 2 mol/L hydrogen chloride, and stir them for 2 h. Eliminate the acidic solution, and wash the resin completely to neutral with distilled water. Add 25 mL of 2 mol/L sodium hydroxide to the above resin and stir for 2 h. Eliminate the basic solution and wash it to neutral with distilled water. Suspense the resin in 50 mL sodium citric acid buffer (pH 4.2), and keep it for use.

(2) Pack the column and equilibrate

Take a chromatography column with the diameter of 0.8-1.2 cm and the length of 10-12 cm, put a piece of cotton or sponge circular cushion at the column bottom, and pour the above-prepared resin from the top. Close the outlet of the chromatography column. When the resin sediment appears, let out the excessive solution, add more resin until the height of the resin sediment is 8-10 cm.

Add pH 4.2 citrate buffer from the top to wash it until the effluent reaches pH 4.2. Close the outlet of the column and keep the level of the liquid surface about 1 cm higher than that of the resin.

(3) Load, elute and collect the eluent

Open the outlet to make the buffer flowing out. When the level of the liquid is at almost the same height with that of resin, close the outlet (don't make the surface of resin dry). Use a long dropper to add 15 drops of mixture of amino acid to the top of resin directly and carefully, and open the outlet to make it flow into the column slowly. When the level of the liquid is just at the same height as that of resin, add 3 mL of 0.1 mol/L hydrogen chloride, and elute it at the flowing rate of 10-12 drops/min. Collect the eluent, 20 drops per tube and collect it one by one. When the level of hydrogen chloride solution is just at the same height as that of resin, use 1 mL of pH 4.2 citric acid buffer to wash the column wall once, then elute with 2 mL of pH 4.2 citric acid buffer. Keep the flowing rate at 10-12 drops/min. Pay attention not running out of the elute.

After eluting out the second amino acid with pH 4.2 critic acid buffer, collect two tubes of the negative parts of the action of triketohydrindene hydrate. Close the outlet of the chromatography column, and transfer the pH 4.2 critic acid buffer left on the top of the resin.

Add 2 mL of 0.1 mol/L sodium hydroxide from the top of the resin; open the outlet to make it flow into the column slowly. Elute with 0.1 mol/L sodium hydroxide according to the above operation, and collect it tube by tube (caution to keep the flowing rate of 10-12 drops/min) by 20 drops per tube. Check up the amino acid in the eluent. After the third amino acid has been eluted out by 0.1 mol/L sodium hydroxide, keep on collecting two tubes of the negative parts of the action of triketohydrindene hydrate.

(4) Monitoring

While collecting the eluent, check the amino acid in the eluent with triketohydrindene tube by tube. The process is add 10 drops of pH 5 acetate acid buffer and 10 drops of neutral triketohydrindene hydrate solution into every tube of eluent. The mixture is heated for 10 minutes in the boiling water bath. The solution showing royal blue means some amino acid has been eluted out. The degree of the color can show the concentration of amino acid, and it can be determined by color matching.

(5) Determining

Similar with monitoring process, detecting the absorption of each tube of eluent at the wavelength of

570 nm and using water as blank as ordinate ( or the degree of color expressed with "−, +, +, ++ ...") , and the tube number of eluent as abscissa, then draw an elution curve.

# Experiment 4   Gel Filtration Chromatography of Hemoglobin and DNP–glutamic Acid

## 4.1   Principle

Gel chromatography, also known as molecular sieve chromatography, is a method used to separate soluble proteins based on different molecular size and shape. Details about gel chromatography see Chapter 5.

Molecules that are larger than the size of pores on the gel are unable to diffuse into the gel and can only pass along the space between the matrix. Molecules that are relatively smaller will enter the pore, which means they will take more time along the long and circuitous route inside the beads. Molecules that range in sizes between the very big from the gel and the very small that can penetrate the pores vary in degrees based on their sizes.

## 4.2   Materials

### 4.2.1   Reagents and solutions

(1) Sephadex G–75: it can separate molecules weight from 3,000 to 70,000.

(2) 0.05 mol/L Tris–HCl: tris base 7.882 g, qs to 1 L with $H_2O$, pH 7.5.

(3) Mixed sample solution: 6 mg hemoglobin and 6 mg DNP ( Dinitrophenyl) –glutamic acid, dissolved in 1 mL 0.05 mol/L Tris–HCl.

### 4.2.2   Special equipment

HD–88–5AI spectrophotometer.

## 4.3   Methods

(1) Sephadex G75 need to be immersed in Tris–HCl for at least 6 h so that it will be swollen.

(2) Gels are poured into the column, and it needs about 1 h to make balance ( always make sure the buffer surface is higher than the matrix and no bubble inside) .

(3) Open the spectrophotometer and set the absorbance work at 280 nm.

(4) Keep washing the column with Tris–HCl for another 60 min until the data on the screen become stable.

(5) Add the mixed sample to the top of the resin by allowing the solution to gently run down.

(6) Place the effluent tube in the first test tube rack and open the clamp.

(7) Run the column at 0.5 mL/min.

(8) Collect the solution and make notes.

*Li Lin*

# Chapter 10

## Separation, Purification, and Identification of Serum Proteins

The separation and purification of proteins are a series of processes intended to isolate one or a few proteins from a complex mixture. It is vital for the characterization of the chemical composition, structure, biological function, and interactions of the proteins of interest. The approaches usually utilize differences in protein physicochemical properties, size, biological activity, and binding affinity.

## Experiment 1 Separation, Purification, and Quantification of Serum Albumin and $\gamma$-globulins

## 1.1 Principle

The separation and purification of proteins are important means to study the chemical composition, structure and biological function of proteins. Its basic principles are based on the properties of proteins. Since the molecular weight, solubility, and isoelectric point of different proteins in human serum are all different, which makes it possible to isolate and purify the proteins in the serum.

### 1.1.1 Salting out

The proteins form stable colloidal solutions of the type known as hydrophilic colloids. Such systems have two stability factors, charge and hydration, either of which is capable of keeping protein molecules in solution. The high concentrations of neutral salts ( such as ammonium sulfate, sodium sulfate, or sodium chloride) can not only remove the shell of water but also can suppress the charge on the surface of the protein molecules, resulting in protein aggregation and precipitation. This phenomenon is referred to as salting out.

The size of particles, the number of charges and the degree of hydrophilicity of various proteins in the serum are all different. The addition of a neutral salt in the right amount can selectively precipitate some proteins, while others remain in solution. For example, serum globulins are precipitated by half-saturated ammonium sulfate, whereas albumins are precipitated only on full saturation. Different proteins in the serum can be separated and purified by using different concentrations of ammonium sulfate solution.

## 1.1.2    Desalination

Salting out is a highly-soluble and normally non-denaturing process. It just changes the shell of water and the charge on the surface of the protein, but can not change the internal structure of the protein which still has all the own natural properties. The protein separated by salting out contains a large number of neutral salts, which will hinder the further purification of the protein. Therefore, desalting is necessary.

In this experiment, the albumin solution or globulin solution containing a large amount of ammonium sulfate passes through a Sephadex G-25 column, then the column is eluted with 0. 02 mol/L $NH_4Ac$ buffer. Because of the differences in molecular size, globulin or albumin ( protein detected with sulfosalicylic acid) and ammonium sulfate ( $SO_4^{2-}$ detected with $BaCl_2$ ) will be successively eluted from Sephadex G-25 column by 0. 02 mol/L $NH_4Ac$ buffer, to achieve desalination of globulin or albumin.

## 1.1.3    Purification of γ-globulin or albumin

The serum protein isolated by salting out is of low purity and requires further purification. The protein is an amphoteric electrolyte. Based on the different isoelectric point of each protein in serum, they can be further separated from each other by ion-exchange chromatography.

Ion-exchange chromatography separates molecules based on their respective charged groups. The charged molecules from the mobile phase bind to the stationary phase carrying an opposite charge to the molecule. Proteins contain charged groups on the surface that interact with the immobilized ions on the stationary phase. In this experiment, diethyl aminoethyl ( DEAE) cellulose anion exchanger is used to purify γ-globulin or albumin. γ-globulin is eluted and separated from albumin, α-globulin, and β-globulin in 0. 02 mol/L $NH_4Ac$ pH 6. 5 elution buffer. Albumin is then eluted and separated from α-globin and β-globin by increasing the ionic strength of the mobile phase (0. 3 mol/L $NH_4Ac$ pH 6. 5). The isoelectric point of serum proteins is shown in Table 10. 1.

Table 10. 1    Isoelectric points, average molecular weights and normal contents of serum proteins

|  | Albumin | Globulin | | | |
|---|---|---|---|---|---|
|  |  | $\alpha_1$ | $\alpha_2$ | β | γ |
| Isoelectric point | 4. 88 | 5. 06 | 5. 06 | 5. 12 | 6. 85–7. 30 |
| Molecular weight ( $\times 10^4$) | 6. 9 | 20 | 30 | 9–15 | 15. 6–30. 0 |
| Content ( %) | 57–67 | 2–5 | 4–9 | 6. 2–12. 0 | 12–20 |

## 1.1.4    Identification of γ-globulin or albumin

Cellulose acetate electrophoresis is a process by which charged particles are separated in a cellulose acetate membrane according to the differences in their charge density. Following staining, the particles are then detected on a cellulose acetate sheet. This method has such virtue as that it can separate the molecules quickly in a small amount of sample. Serum proteins are negatively charged when migrating in barbital buffer of pH 8. 6 because all of the pI values are lower than pH 7. 0.

## 1.1.5    Determination of γ-globulin or albumin

Proteins produce a color change in certain reactions. Due to characteristic groups of particular amino acids present in it, these reactions are not quite specific for a protein molecule. The biuret reaction is a method that is normally used to identify the presence of certain proteins and determine their concentration.

When heated, urea forms biuret. If a strongly alkaline solution of biuret is heated together with very dilute copper sulfate, a purple-blue color is obtained. The biuret reaction occurs with all compounds containing two or more peptide bonds. The name of the reaction is derived from the organic compound biuret.

The color intensity of this reaction varies in proportion with the concentration of the protein. The spectrophotometer can then be used to measure the intensity of the color for the quantification of the protein.

# 1.2　Materials

## 1.2.1　Reagents and solutions

(1) The sample solution ( Blood serum: fresh healthy human serum or animal serum, no hemolysis, no sediment or bacteria breeding).

(2) Saturated ammonium sulfate solution: weigh 850 g solid ammonium sulfate( $NH_4$ )$_2SO_4$ and dissolve in 1,000 mL distilled water with agitating for solubilization at 70–80 ℃. The ammonium sulfate solution is placed overnight at room temperature. When the white crystallization is precipitated at the bottom of the bottle, the supernatant is a saturated ammonium sulfate solution.

(3) 0.3 mol/L pH 6.5 $NH_4$Ac buffer: weigh 23.12 g ammonium acetate( $NH_4$Ac), dissolve in 800 mL distilled water, adjust its pH to 6.5 with dilute ammonia or dilute acetic acid, and the buffer is further diluted to 1,000 mL with distilled water.

(4) 0.06 mol/L pH 6.5 $NH_4$Ac buffer: the above buffer is diluted by 1 : 5 in distilled water.

(5) 0.02 mol/L pH 6.5 $NH_4$Ac buffer: the above buffer is diluted by 1 : 3 in distilled water.

(6) 1.5 mol/L NaCl–0.3 mol/L $NH_4$Ac buffer: weigh 87.7 g sodium chloride ( NaCl), dissolve in 0.3 mol/L pH 6.5 $NH_4$Ac buffer, and add distilled water to 1,000 mL.

(7) 0.92 mol/L (20%) sulfosalicylic acid solution.

(8) 0.05 mol/L (1%) $BaCl_2$ solution.

(9) Biuret reagent: weigh 3.0 g cupric sulfate ( $CuSO_4 \cdot 5H_2O$) and 9.0 g potassium sodium tartrate ( $NaKC_4H_4O_6 \cdot 4H_2O$) and 5.0 g of potassium iodide( KI); dissolve separately in about 25 mL distilled water. Transfer sodium potassium tartrate solution and potassium iodide solution to a dry 1 L volumetric flask, and add 100 mL of 6.0 mol/L sodium hydroxide ( NaOH) into the volumetric flask, mix thoroughly. Keep stirring while adding the copper sulfate solution. The solution is further diluted by distilled water to 1,000 mL. Mix the contents thoroughly and store in a plastic bottle.

(10) Standard protein solution (5.0 mg/mL BSA).

(11) 0.9% NaCl solution.

(12) Barbiturate buffer ( pH 8.6, 0.07 mol/L, ionic strength 0.06).

(13) 0.5% amino black 10B staining solution.

(14) Destaining solution: alcohol 45 mL, glacial acetic acid 5 mL, qs to 100 mL with Milli–Q $H_2O$.

## 1.2.2　Special equipment

①Sephadex G–25 chromatography column ( 1.0×10 cm); ②DEAE–cellulose chromatography column (1.0×6 cm); ③table centrifuge; ④electrophoresis equipment; ⑤power supply; ⑥micropipettes; ⑦microfuge tubes; ⑧spectrophotometer.

# 1.3　Methods

## 1.3.1　Separation of γ–globulin or albumin

(1) Salting out

1) Add 0.8 mL blood serum into a centrifuge tube. While stirring, add 0.8 mL the saturated ammoni-

um sulfate solution drop-wise to the centrifuge tube.

2) Mixed thoroughly, the tube is placed stably for 10 min at room temperature, centrifuge at 4,000 r/min for 10 min.

3) Transfer the aqueous phase to a fresh tube for purification of albumin.

4) Add 0.6 mL distilled water into the precipitation in the centrifuge tube. Dissolve the precipitation by shaking for purification of γ-globulin.

(2) Desalination by gel chromatography

1) Preparation of the gel

● Weigh 25.25 g Sephadex G-25 per 100 mL column volume, add distilled water about 50 mL/g dry gel. Incubate in boiling water bath 1 hour and then remove it for cooling. Discard the supernatant containing fine suspended solids after gel sinking. And repeat the process 2-3 times.

● Then repeat the following operations twice: add 2 volumes of 0.02 mol/L pH 6.5 $NH_4Ac$ buffer, gently stirring, and place a moment. Discard the supernatant after cellulose sinking.

2) Packing the column

● Select a 1.0 cm×10 cm chromatography column, fix it on the holder, and keep it vertical. Close the outlet of the column. Add a small amount of 0.02 mol/L pH 6.5 $NH_4Ac$ buffer to the bottom of the column.

● Gently stir the treated gel particle suspension with a glass rod to create a slurry. Fill the column carefully to avoid the introduction of air bubbles as much as possible to obtain gel homogeneity in the column. The evenest packing can be achieved by pouring all slurry into the column at once. If necessary, stir the settling gel to prevent the formation of gel layers.

● Open the outlet of the column at the bottom and allow the column to pack under gravity until the proper height of the column bed (about 7 cm) is formed.

● Add the buffer as the column bed is packed. Do not let the buffer drop below the top of the column bed.

● Close the outlet of the column when the buffer surface is close to the top of the cellulose column bed.

3) Equilibration

The column is equilibrated with the buffer for the separation.

● Carefully adjust the clamp of the outlet tube of the column to allow the flow rate to be 2 mL/min (40 drops/min).

● Equilibrate the column with 0.02 mol/L pH 6.5 $NH_4Ac$ buffer for at least 20-30 min. Close the outlet of the column, then the column is ready for the use of desalting at this time.

4) Loading sample

Loading sample is the most important procedure. The more evenly it is loaded, the better the result will be obtained. For the best resolution, the sample volume should not exceed 1%-2% of the column volume.

● Carefully adjust the clamp of the outlet tube of the column. Allow the buffer to drain onto the gel until the surface of the bed is just barely dry, and then close the outlet of the column. Never permit the top of the column to dry out.

● Load the crude protein sample solution (approximately 0.8 mL albumin or globulin solution) with a pipette gently to the top of the column and a 10 mL graduated centrifuge tube is used to collect the eluent (in order to know the outflow after the sample is loaded).

● Adjust the clamp to allow a flow rate of 2 mL/min (40 drops/min). Let the sample all run in, and add 1 mL of 0.02 mol/L pH 6.5 $NH_4Ac$ buffer to clean the protein samples that stick to the wall of the column.

5) Elution

● Add additional 0.02 mol/L pH 6.5 $NH_4Ac$ buffer for a continuous elution. Pay attention to the amount of effluent at the same time.

6) Detection and collection of proteins

• Add 2 drops well of 0.92 mol/L sulfosalicylic acid in the wells of the black reaction plate. Add 1 drop of eluent to the well containing the sulfosalicylic acid solution. The occurrence of white turbidity or precipitation indicates that the eluent contains protein. If the gel bed volume is 5.5 mL, there is a possibility that the protein flows out when the amount of the outflow liquid is about 2 mL.

• Immediately start collecting the eluent containing the protein. After collecting about 12 drops, add one drop of eluate to the well of a black reaction plate pre-added 0.05 mol/L $BaCl_2$. When white precipitate appears (indicating $SO_4^{2-}$), stop the collection immediately. The collected protein solution can be further purified using DEAE cellulose ion-exchange chromatography.

7) Regeneration and preservation

• After completing the collection of the protein solution, the gel column can be further eluted with 0.02 mol/L pH 6.5 $NH_4Ac$ buffer until the $BaCl_2$ detection of eluent is $SO_4^{2-}$ negative. At this point, continue to elute with 2-3 mL of 0.02 mol/L $NH_4Ac$ buffer. To reuse the column, it should be equilibrated with the required buffer solution after each use.

8) If it is not used for the time being, the gel column should be washed with a buffer containing 3 mmol/L (0.02%) $NaN_3$, then stored in the buffer to prevent mildew. If the column is not used for a long time, the gel should be poured out from the column and add $NaN_3$ to 3 mmol/L and stored in a wet state at 4 °C.

## 1.3.2　Purification of γ-globulin or albumin

(1) Purification of γ-globulin

1) Preparation of DEAE-C

• Weigh 14 g DEAE-C per 100 mL column volume, add 0.5 mol/L HCl to DEAE-C with a ratio of 15 : 1 (15 mL HCl/g DEAE-C). Mix well by stirring and place for 30 min. Add 10 times the amount of distilled water, stirring evenly, place a moment. Discard the supernatant containing fine suspended solids after cellulose sinking. And repeat this process 2-3 times.

• Place DEAE-C into a brucella funnel with a fine nylon filter cloth and wash fully with distilled water until the effluent pH is about 4.

• Transfer the DEAE-C into a beaker, and add 0.5 mol/L NaOH into the beaker. The DEAE-C is repeated in the above process until the effluent pH is about 7.

2) Packing the column

• Place DEAE cellulose treated as described above in a beaker. Add 0.02 mol/L pH 6.5 $NH_4Ac$ buffer with a ratio of 40 : 1 (40 mL $NH_4Ac$/g cellulose) to the beaker. And adjust pH to 6.5 with acetic acid. Discard the supernatant and the DEAE cellulose is ready for packing the column.

• Close the outlet of the column. Add a small amount of buffer to the bottom of the column.

• Stir the DEAE cellulose to create a slurry and carefully fill the column, avoiding the introduction of air bubbles as much as possible to keep gel homogeneity.

• Open the outlet of the column at the bottom and allow the column to pack under gravity until the proper height of the column bed (about 6 cm).

• Add the buffer as the DEAE cellulose packs. Do not let the buffer drop below the top of the column bed!

• Close the outlet of the column when the buffer surface is close to the top of the cellulose column bed.

3) Equilibration

The column is equilibrated with the buffer for the separation.

• Equilibrate the column with several volumes of 0.02 mol/L pH 6.5 $NH_4Ac$ buffer for at least

30 min.

4) Loading sample

● Carefully adjust the clamp of the outlet tube of the column. Allow the buffer to drain onto the DEAE−C surface until the surface of the bed is just barely dry, and then close the outlet of the column. Never permit the top of the column to dry out.

● Load the sample with a pipette gently to the top of the column. Adjust the clamp and let the sample run into the surface. Allow all to run in, and add 1 mL 0. 02 mol/L pH 6. 5 $NH_4Ac$ buffer to clean the protein samples that stick to the wall of the column.

● Immediately start collecting the eluent in a 10 mL scale centrifuge tube as soon as you let the sample run into the gel.

5) Elution

● Add additional 0. 02 mol/L pH 6. 5 $NH_4Ac$ buffer for a continuous elution. Pay attention to the amount of effluent at the same time.

6) Detection and collection of γ−globulin

● Detect the protein in eluent by 0. 92 mol/L (20%) sulfosalicylic acid at any times.

● When the protein appears in the eluent, collect 3 tubes of the eluent continuously, 10 drops/tube, and a tube with the highest protein concentration is prepared for the purity analysis and identification.

(2) Purification of albumin

1) Loading sample

● Carefully adjust the clamp of the outlet tube of the column. Allow the buffer to drain to the DEAE−C surface. Close the outlet of the column when the surface of the bed is just barely dry. Never permit the top of the column to dry out.

● Add slowly the desalted albumin solution with a pipette to the surface of the column bed. Adjust the clamp and let the sample runinto the surface. Allow all run in, add 1 mL of 0. 06 mol/L pH 6. 5 $NH_4Ac$ buffer to clean the protein samples that stick to the wall of the column.

● Immediately start collecting the eluent in a 10 mL scale centrifuge tube as soon as the sample run into the gel.

2) Elution

● Add additional 0. 06 mol/L pH 6. 5 $NH_4Ac$ buffer for a continuous elution. Pay attention to the amount of effluent at the same time.

● After collecting about 6 mL effluent (including α−globulin and β−globulin), allow the buffer to drain to the column bed until the surface of the bed is just barely dry. Add 0. 3 mol/L pH 6. 5 $NH_4Ac$ buffer for a continuous elution.

3) Detection and collection of albumin

● Detect the protein in eluent by 0. 92 mol/L (20%) sulfosalicylic acid. The purified albumin is still combined with a small amount of bile pigment. It can be observed as the sample makes its way down the column.

● When the 0. 3 mol/L $NH_4Ac$ buffer is eluted for about 2. 5 mL, collect immediately 2 tubes of the eluent continuously, 10 drops/tube, and the eluents are ready for the purity analysis and identification of albumin.

4) Regeneration and preservation

● Wash the used chromatography column by the 1. 5 mol/L NaCl−0. 3 mol/L $NH_4Ac$ buffer, and then by about 10 mL 0. 02 mol/L pH 6. 5 $NH_4Ac$ buffer. It is ready to be reused.

● DEAE cellulose should be stored in a buffer containing 0. 11 mol/L (1%) n−butanol in order to prevent mildew if it is not used for the time being.

## 1.3.3   Identification of γ-globulin or albumin

(1) Preparation of CAM

● Prepare four of 2 cm×8 cm cellulose acetate membranes.

● Make a marker and draw a line gently with pencil on the smooth side of the membrane, about 1. 5 cm far away from one short edge.

● Soak the membranes in the barbital buffer ( pH 8. 6) for about 20 min, with smooth side upward. Take out CAM and clip it in clean filter paper, gently suck off the excess buffer on the surface.

(2) Loading sample

● Drop a gutta of the sample ( e. g. , blood serum, purified γ-globulin, purified albumin) on the glass slide.

● Stain the sample with one short edge of a cover glass or a sampler.

● Print the sample gently on the back of the pencil line ( on the rough side).

(3) Electrophoresis

● Put the membrane on the paper bridge in the electrophoresis tanker with rough side downward and sample near the cathode.

● After balancing for 5-10 min, start the electrophoresis. Stop after migrating 45-60 min, at 110-130 V.

(4) Staining

● Soak the membranes in 0. 5% amino black 10B staining solution for 5-10 min with, rough side upward.

(5) Rinsing

● Remove the membranes from the dye and wash the excess dye with tap water.

● Rinse them 3-4 times in destaining solution for 10 min, for every time, take out and dry up with filter papers.

## 1.3.4   Determination of γ-globulin or albumin

(1) Preparation of samples: the normal serum is diluted by 1 : 10 with 0. 9% sodium chloride. The diluted normal serum is for sample 1. Purified γ-globulin is for sample 2. Purified albumin is for sample 3.

(2) Prepare 5 test tubes and operate according to Table 10. 2.

Table 10. 2   Determination of protein content in each sample by Biuret assay

|  | Reagent blank ( mL) | Standard ( mL) | Sample 1 ( mL) | Sample 2 ( mL) | Sample 3 ( mL) |
|---|---|---|---|---|---|
| 1 : 10 Serum/plasma | — | — | 0. 5 | — | — |
| γ-Globulin solution | — | — | — | 0. 5 | — |
| Albumin solution |  |  |  |  | 0. 5 |
| Standard protein solution | — | 0. 5 | — | — | — |
| 0. 9% NaCl | 1. 0 | 0. 5 | 0. 5 | 0. 5 | 0. 5 |
| Biuret reagent | 2. 0 | 2. 0 | 2. 0 | 2. 0 | 2. 0 |

Mix well and incubate at 37 ℃ for 20 min.

(3) Determination: measure the absorbance of the standard and the sample against the reagent blank by spectrophotometer at 540 nm wavelength.

# 1.4 Results and discussion

(1) Results of cellulose acetate electrophoresis: the stained cellulose acetate membrane is placed side by side while the sample line is at the same level. Electrophoretic diagram of normal serum protein as a control. Observe, analyze and describe the type and purity of the serum protein separated in the experiment.

The normal human serum protein should be separated into 5 bands: alumni, $\alpha_1$ –globular protein, $\alpha_2$ – globular protein, $\beta$ –globular protein, and $\gamma$ – globular protein, just as the following Figure 10.1.

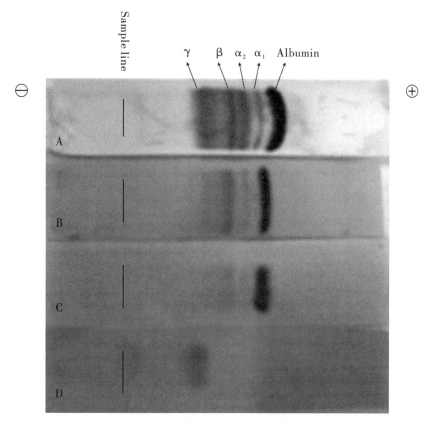

A. Normal serum; B. Purified albumin in the first eluent; C. Purified albumin in the second eluent; D. Purified $\gamma$–globulin.

Figure 10.1 Electrophorogram of $\gamma$–globulin or albumin

(2) Results of determination of $\gamma$–globulin or albumin( Table 10.3).

Table 10.3 Original records of quantitative determination of protein sample

|  | Standard $(A_s)$ | Sample 1 $(A_{u1})$ | Sample 2 $(A_{u2})$ | Sample 3 $(A_{u3})$ |
|---|---|---|---|---|
| 1 |  |  |  |  |
| 2 |  |  |  |  |
| 3 |  |  |  |  |
| $\overline{A_{540}}$ |  |  |  |  |

(3) Calculation: calculation of the protein concentration according to the following formula.

$$\text{Blood serum protein(mg/mL)} = \frac{A_{u1}}{As} \times 5.0 \text{ (mg/mL)} \times \frac{1 \text{ mL}}{0.1 \text{ mL}} = \frac{A_{u1}}{As} \times 5.0(\text{mg/mL}) \times 10$$

$$\gamma\text{-Globulin(mg/mL)} = \frac{A_{u2}}{As} \times 5.0(\text{mg/mL})$$

$$\text{Albumin(mg/mL)} = \frac{A_{u3}}{As} \times 5.0(\text{mg/mL})$$

(4) Purities of serum γ-globulin and albumin separated in the experiment are identified by cellulose acetate membrane electrophoresis. What is the basis by which you estimate their purity and variety?

(5) What is the principle of separation and purification of proteins by Sephadex chromatography? Why can γ-globulin be separated from ammonium sulfate by Sephadex G-25?

*Zhang Weijuan*

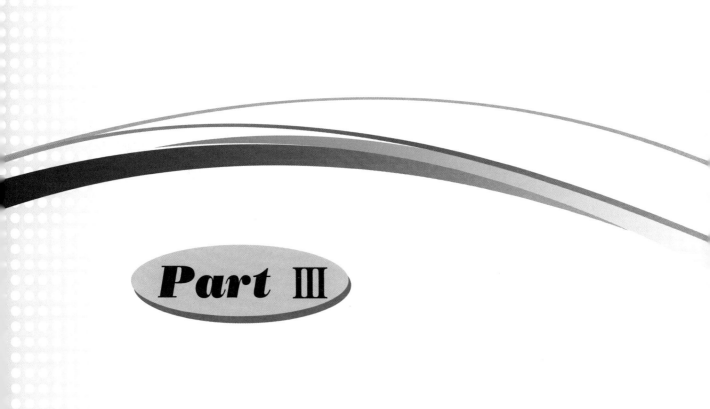

# Part III

# Enzymology Experiments

# Chapter 11

## Determination of Enzyme Activity

Enzyme activity is the ability of an enzyme to catalyze a specific reaction, which can be determined by the changes in the concentration of a reaction product or reactant in a certain time. The higher the rate of the enzymatic reaction, the greater the enzyme activity. The enzyme activity was expressed in terms of enzyme activity units (U). In 1961, the international society of biochemistry enzymology committee proposed a unified international unit (IU) to represent the activity of the enzyme, which is defined as: under the optimum conditions (25 ℃), the amount of enzyme required to catalyze the conversion of 1 micromolar (mmol) substrates to products per minute is defined as one unit of activity.

There are three common strategies for determining the rate of enzymatic reaction. ① Mix the enzyme and its substrates under appropriate conditions, let the mixture react for a certain time, then stop the reaction and determine the amount of substrate reduction or product increase. ②The enzyme and substrates are mixed to determine the time required to produce a certain amount of products or to consume a certain amount of substrates. ③After the enzyme and substrates are mixed, the changes of reaction system, such as the absorbance increase or decrease, are determined intermittently or continuously.

The enzyme is an important part of medical research. It is also the content of analysis and detection in clinical practice.

## Experiment 1    Determination of Alanine Aminotransferase Activity in Serum

## 1.1    Principle

Serum alanine aminotransferase (ALT) activity has been regarded as a reliable and sensitive marker for the course of liver damage in clinical settings and in safety assessment of pharmaceuticals and chemicals.

Many methods are used to determine transaminase activity. In this experiment, the activity of serum ALT is determined colorimetrically by modified Lai's method. Alanine aminotransferase catalyzes the production of pyruvate and glutamic acid by transferring an amino group from alanine to alpha-ketoglutaric acid. The amount of pyruvate production represents the magnitude of reactive enzyme activity. Pyruvate and 2,4-dinitrophenylhydrazine in the acidic solution can form yellow 2,4-dinitrophenylhydrazone, which is red−brown in alkaline condition (Figure 11. 1). By measuring the absorbance of 2, 4−dinitrophenylhydrazone at 520 nm, the ALT activity is determined in serum, so this method is also called determination colorimetric method. Although this method has shortcomings, the operation is simple, and do not need special equipment and reagents, thus it is widely used in the clinic.

$$\text{alanine + alpha ketoglutaric acid} \underset{\text{}}{\overset{\text{ALT}}{\rightleftharpoons}} \text{pyruvic acid + glutamic acid}$$

$$\text{pyruvic acid + 2,4−dinitrophenylhydrazine} \longrightarrow \text{2,4−dinitrophenylhydrazone(yellow)}$$

$$\text{OH}^- \downarrow \searrow \text{H}_2\text{O}$$

phenylhydrazone nitraquinone compound

(red brown)

Figure 11.1   The catalytic reaction of alanine aminotransferase

# 1.2   Materials

## 1.2.1   Reagents and solutions

(1) Sample: human serum.

(2) 0.1 mol/L phosphate buffer solution (pH 7.4).

(3) 2.0 mol/mL sodium pyruvate standard solution: dissolve 11 mg purified sodium pyruvate in 50 mL phosphate buffer solution(prepare on the same day).

(4) Alanine aminotransferase substrate: add 1.78 g L−alanine and 29.2 mg analytically pure alpha−ketoglutarate in a small beaker, and use 1 mol/L hydrochloric acid (HCl) or 1 mol/L sodium hydroxide (NaOH) solution to adjust pH to 7.4, and then add phosphate buffer to 100 mL. Add a few drops of chloroform to prevent corruption. This solution contains 2 mol/L alpha−ketoglutaric acid and 0.2 mol/L of alanine (This solution can be stored in the refrigerator for one week).

(5) 2, 4−dinitrophenylhydrazine solution: put 19.8 mg analytically pure 2, 4−picrylhydrazyl in the 200 mL conical flask, and then add 100 mL of 1 mol/L hydrochloric acid. The conical flask is placed in the dark, and continue to shake it until 2, 4−dinitrophenylhydrazine was dissolved. Filter the solution into the brown bottle and keep it in the refrigerator.

(6) 0.4 mol/L sodium hydroxide solution.

## 1.2.2   Special equipment

①Spectrophotometer; ②acidometer; ③volumetric flask.

# 1.3   Methods

## 1.3.1   Plot the standard curves

(1) Label 5 test tubes as 1−5. Add the reagents according to the order listed in the following Table 11.1.

Table 11.1 Reaction system

| Reagent (mL) | 1 | 2 | 3 | 4 | 5 |
|---|---|---|---|---|---|
| Standard sodium pyruvate solution | 0 | 0.05 | 0.10 | 0.15 | 0.20 |
| Alanine aminotransferase substrate | 0.5 | 0.45 | 0.40 | 0.35 | 0.30 |
| Phosphate buffer solution (0.1 mol/L, pH 7.4) | 0.10 | 0.10 | 0.10 | 0.10 | 0.10 |
| The corresponding amount of Carmen's enzyme activity units * | 0 | 28 | 57 | 97 | 150 |

* NADH is oxidized to $NAD^+$, and the rate of NADH decrease (photometrically determined at 340 nm) is directly proportional to the rate of formation of pyruvate and the ALT activity. Thus, Carmen's unit is defined as the amount of produced pyruvic acid catalyzed by 1 mL of serum (3 mL of the total reaction solution) at 340 nm and 25 ℃ in one minute, which then reacts with reduced coenzyme Ⅰ by the catalysis of lactate dehydrogenase to produce oxidized coenzyme Ⅰ, decreasing the absorbance to 0.001. This is one unit of transaminase activity.

(2) Place the test tube in a constant temperature water bath at 37 ℃ for 10 min to balance the temperature inside and outside.

(3) Add 0.5 mL of 2,4-dinitrophenylhydrazine solution to each tube, mix and incubate them at 37 ℃ for 20 min.

(4) Add 5 mL of 0.4 mol/L sodium hydroxide solution into each tube, then mix and stay at room temperature for 15 min.

(5) Use distilled water as blank, determine the light absorption values at 520 nm in tube 1-5 with spectrophotometry. Draw the standard curve by using the difference value that is gotten by subtracting the absorbance of tube 1 from the absorbance of tube 2, 3, 4, 5, respectively, as a vertical coordinate, and the number of Carmen's activity unit as abscissa.

### 1.3.2 Determine enzyme activity

(1) Label 2 tubes as a test tube and a blank tube for control.

(2) Add 0.5 mL alanine aminotransferase substrate into each tube, and place them in 37 ℃ water bath for 10 min, so that the temperature balance inside and outside of the tubes. Then add 0.1 mL serum to test tube, and keep these two tubes in the 37 ℃ water bath for 30 min.

(3) Add 0.5 mL 2,4-dinitrophenylhydrazine reagent into each of the two tubes, and keep them in the 37 ℃ water bath for 20 min.

(4) Add 5 mL of 0.4 mol/L sodium hydroxide solution into each tube, and stay at room temperature for 15 min.

(5) Use distilled water as blank, determine the absorbance of the test tube at 520 nm with spectrophotometry. Calculate the absorption value of the test tube according to the method of step 5, and find the unit of activity of transaminase in each 100 mL serum on the standard curve.

## 1.4 Results and discussion

(1) An example of calculating serum alanine aminotransferase activity by using a standard curve of pyruvate in modified Lai's method.

Five pyruvate standard samples of different Carmen's active units, such as 0, 28, 57, 97, and 150, were set up. The absorbance was measured by Lai's method. The standard curve was fitted with Carmen's enzyme activity unit (U) as the horizontal coordinate and the corresponding absorbance as the ordinate (Figure 11.2).

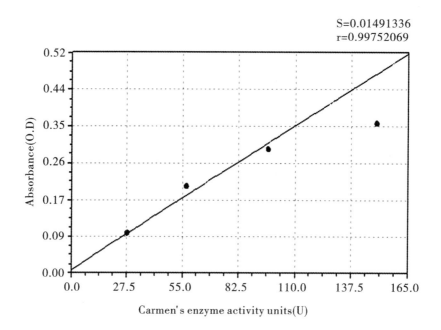

S=0.01491336
r=0.99752069

Figure 11.2  **The standard curve**

If the $A_{520}$ value of the test tube—the $A_{520}$ value of the blank tube =0.17, then you can find that the difference of 0.17 corresponds to 55 Carmen's enzyme activity units in serum for alanine aminotransferase in the standard curve.

(2) What does the test result mean?

Low levels of ALT(2–40 U) in the human blood are expected and are normal. Very high levels of ALT (10 times more than normal) are usually due to acute hepatitis, sometimes due to a viral infection. With acute hepatitis, ALT levels usually stay high for 1–2 months but can take as long as 3–6 months to return to normal. Levels of ALT may also be markedly elevated (often over 100 times than normal) as a result of exposure to drugs or other substances that are toxic to the liver as well as in conditions that cause decreased blood flow (ischemia) to the liver. Pregnancy, a shot or injection of medicine into muscle tissue, or even strenuous exercise may increase ALT levels. Acute burns, surgery, and seizures may raise ALT levels as well.

# Experiment 2   Determination of Salivary Amylase Activity

## 2.1   Principle

An amylase belongs to the hydrolase, which is a biological catalyst of starch hydrolysis.

Under the action of salivary amylase, starch is hydrolyzed and passed through a series of intermediate products called dextrin. Finally, maltose and glucose are produced.

Starch ——→purple dextrin ——→red dextrin ——→maltose, glucose.

Starch, purple dextrin and red dextrin react with iodine and show blue, purple and red, respectively. However, maltose and glucose do not react with iodine. Amylase breaks down starch into the maltose as the end product.

In this experiment, the color reaction of iodine is used to determine the rate of hydrolysis of starch by salivary amylase to assay the size of salivary amylase activity. Starch solutions with known concentrations are used as substrates, adding a certain amount of saliva, and the salivary amylase will be part of hydrolysis of starch under suitable conditions. The substrate excess conditions, reaction after adding iodine and starch hydrolysis has not been combined into a blue complex, shades of blue and the blank without enzymatic reaction tube, which calculate amylase content. The purpose of this experiment is to study the enzyme amylase which is found in saliva.

## 2.2　Materials

### 2.2.1　Reagents and solutions

(1) Iodine storage solution: add 11 g iodine and 22 g potassium iodide into 500 mL capacity bottle, then add a little distilled water to make iodine and potassium iodide completely dissolved. Finally, fix the volume to 500 mL, store it in a brown bottle.

(2) Diluted iodine solution: absorb 2 mL the iodine storage solution, add 20 g potassium iodide, then dissolve it with distilled water. Finally, fix the volume to 500 mL, and store it in a brown bottle.

(3) Standard "finish color" solution: ① Accurately weigh 40.243,9 g cobalt chloride and 0.487,8 g potassium dichromate, dissolve them in distilled water and fix volume to 500 mL. ② Accurate weigh 40 mg chrome black T, dissolve them in distilled water and fixed volume to 100 mL, and then get 80 mL ① solution and 10 mL ② solution and mix them thoroughly, which is standard "finish color" solution.

(4) 2% soluble starch solution: weigh 2 g dry soluble starch. Firstly mix it with a little distilled water, then pour it into 80 mL boiling water, and continue to boil until the liquid becomes transparent. After cooling, fix the volume to 100 mL with distilled water( fresh preparation is required).

(5) 100 times diluted saliva: rinse your mouth with distilled water to remove food debris. Then let saliva flow into the cylinder and diluted 100 times, mix and reserve.

(6) 0.02 mol/L, pH 6.8 disodium hydrogen phosphate−citric acid buffer solution: weigh 45.23 g disodium hydrogen phosphate and 8.07 g citric acid, and then fix the volume to 1,000 mL with distilled water. Adjust pH by pH meter or precision test paper.

### 2.2.2　Special equipment

①White porcelain board; ②50 mL triangle bottle; ③100−1,000 mL capacity bottles.

## 2.3　Methods

### 2.3.1　The color reaction of iodine

(1) Drop the standard color solution on white porcelain board in the upper left corner of the hole as a standard for comparing terminal colors ( as shown in Figure 11.3).

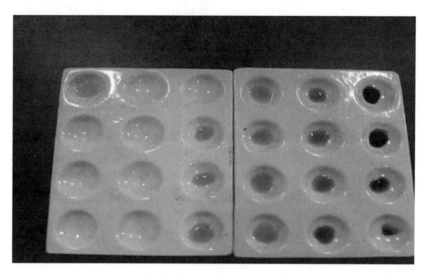

Figure 11.3　The color reaction of iodine

(2) Add 20 mL of 2% soluble starch solution and 5 mL of the 0. 02 mol/L, pH 6. 8 disodium hydrogen phosphate–citric acid buffer solution in a 50 mL triangle bottle. Mix and incubate it at 37 ℃ water bath for 10 min.

(3) Add 0. 5 mL of 100 times diluted saliva, instant recording time and shaking well.

(4) Take out the reaction fluid about 0. 25 mL with transferpettor, and drop it in the hole of white porcelain board which was filled with dilute iodine solution ( about 0. 75 mL) in advance.

(5) When the color of the hole changes from purple to brown–red which is the same as the standard color solution, that is the end point of the reaction, and record the reaction time $T$ ( min).

## 2.3.2　Calculation

(1) Under the condition of 37 ℃ and pH 6. 8, 1 mL of amylase catalyzes the decomposition of soluble starch in 1 hour, which is called the activity unit of salivary amylase.

(2) The formula of the enzyme unit are as follows.

Enzyme–activity unit = $(60/T \times 20 \times 2\% \times n)/0. 5$

$n$——times of dilution of salivary amylase

60——1 h( 60 min)

0. 5——absorb the volume of amylase in saliva to be measured( mL)

2% ——starch concentration

20——the volume of 2% soluble starch solution( mL)

$T$——reaction time( min)

## 2.4　Results and discussion

(1) What does the test result mean?

Salivary amylase, found in humans, is an enzyme that catalyzes the hydrolysis of starch into simpler compounds. Its enzymatic activity is affected by several factors, such as temperature and pH. The rates of enzymatic activity of salivary amylase in different temperatures and pH were measured and resulted to be different. However, due to some errors that were committed, the expected optimum temperature and the optimum pH were not achieved.

This is a relatively simple experiment. However, the activity of human saliva amylase in each person at different times or different conditions is the difference, which leads to different experimental results you got.

(2) Several precautions should be emphasized.

1) All the time should be controlled at 2.0-2.5 min, otherwise, the dilution times should be changed and you had to redetermine.

2) In the experiment, the amount of 2% soluble starch and diluted saliva amylase must be accurate, otherwise, the error will be larger.

3) Every step of the experiment process, the time control should be accurate, the action should be rapid, as far as possible to reduce the interference of environmental changes on the experimental results.

*Wang Huilian*

# Chapter 12

## Analysis of Enzyme Kinetics

Enzymes are protein molecules that act as biological catalysts by increasing the rate of reactions without changing the overall process. Enzymes are seen in all living cells and control the metabolic processes in which they converted nutrients into energy and new cells. Enzymes also help in the breakdown of food materials into its simplest form. The reactants of enzyme–catalyzed reactions are termed as substrates. Each enzyme is quite specific in character, acting on a particular substrate to produce a particular product.

The central approach for studying the mechanism of an enzyme–catalyzed reaction is to determine the rate of the reaction and its changes in response with the changes in parameters such as substrate concentration, enzyme concentration, pH, temperature, etc. This is known as enzyme kinetics. Enzyme kinetics is the study of the chemical reactions that are catalyzed by enzymes.

In enzyme kinetics, the reaction rate is measured and the effects of varying the conditions of the reaction are investigated. Studying an enzyme's kinetics in this way can reveal the catalytic mechanism of this enzyme, its role in metabolism, how its activity is controlled, and how a drug or an agonist might inhibit the enzyme.

## Experiment 1　Determination of $K_m$ of Catalase

## 1.1　Principle

Catalase (hydrogen peroxide oxidoreductase) is an important cellular antioxidant enzyme that defends against oxidative stress.

Catalase is very specific in its reaction, which is the conversion of hydrogen peroxide to water and oxygen. Catalase activity can be measured quantitatively by allowing the enzyme solution to react with hydrogen peroxide for varying periods of time and measuring the excess peroxide remaining by titration with potassium permanganate ($KMnO_4$). This procedure has been found to be practical and rapid. Catalase decomposes hydrogen peroxide ($H_2O_2$) into molecular oxygen and water as shown below:

$$2H_2O_2 \xrightarrow{\text{Catalase}} 2H_2O + O_2$$

In the chemical reaction formula of catalase, the substrate concentration $[S]$ is equal to the concentration of $H_2O_2$ added.

Reaction velocity ($V$) represents the consumed amount of $H_2O_2$ in a fixed time of chemical reaction.

$V$ = added amount of $H_2O_2$ – remained amount of $H_2O_2$.

There mained amount of $H_2O_2$ can be determined by $KMnO_4$ titration with the presence of sulphuric acid.

$$2KMnO_4 + 5H_2O_2 + 3H_2SO_4 \longrightarrow 2MnSO_4 + K_2SO_4 + 5O_2 \uparrow + 8H_2O$$

A Hanes−Woolf plot is used for the evaluation of Michaelis−Menten constants ($K_m$) of the catalase. The Michaelis−Menten equation can be rearranged as follows:

$$\frac{[S]}{v} = \frac{1}{V_{max}}[S] + \frac{K_m}{V_{max}}$$

In Hanes−Woolf plot, the ratio of the initial substrate concentration $[S]$ to the reaction velocity $V$ is plotted against $[S]$. $K_m$ can be determined from the intercept on the $x$−axis ($-K_m$) (Figure 12.1).

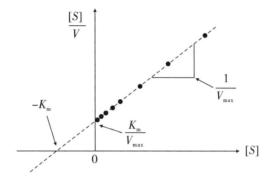

Figure 12.1   Hanes−Woolf plot

# 1.2   Materials

## 1.2.1   Reagents and solutions

(1) 0.02 mol/L phosphate buffer solution (pH 7.4).

(2) 0.02 mol/L $KMnO_4$: dissolve 3.2 g of $KMnO_4$ in distilled water to 1,000 mL and boil for 15 min. After 2 days, filter and store it in a brown bottle.

(3) 0.004 mol/L $KMnO_4$: dissolve 0.2 g constant sodium oxalate in 250 mL of cold boiling water and 10 mL of concentrated sulfuric acid, then titrate to reddish color with 0.02 mol/L $KMnO_4$, heat to 65 ℃ and continue titrating until red does not go back. Calculate the exact concentration of $KMnO_4$ and dilute it to 0.004 mol/L.

$$5Na_2C_2O_4 + 2\ KMnO_4 + 8\ H_2SO_4 = 10\ CO_2 \uparrow + 2\ MnSO_4 + K_2SO_4 + 8H_2O + 5Na_2SO_4$$

(4) 0.05 mol/L $H_2O_2$: add 23 mL of 30% $H_2O_2$ in distilled water to 1,000 mL, calibrate the concentration with 0.004 mol/L $KMnO_4$(dilute 4 times before calibration, then take 2 mL of diluted solution and add 2 mL of 25% $H_2SO_4$. Finally, titrate with 0.004 mol/L $KMnO_4$ to reddish color, after calibration, the concentration of $H_2O_2$ was in fact 0.053 mol/L).

(5) 25% $H_2SO_4$.

## 1.2.2   Special equipment

①50 mL triangle bottle; ②500−1,000 mL capacity bottle.

# 1.3 Methods

## 1.3.1 Calibration of $H_2O_2$ concentration

(1) Hemodilution: dilute 0.1 mL fresh (or treated with heparin) blood to 10 mL with distilled water, and mix thoroughly. Take 1.0 mL this diluted blood and dilute it to 10 mL with phosphate buffer (pH 7.0, 0.2 mol/L), then obtain 1 : 1,000 diluted blood.

(2) Calibration of $H_2O_2$ concentration: take two clean conical flasks. Add 2.0 mL of 0.08 mol/L $H_2O_2$ and 2.0 mL of 25% $H_2SO_4$, respectively (high concentration of sulfuric acid, be careful not to spill). Then titrate them with 0.004 mol/L of $KMnO_4$ to reddish color, respectively. Therefore, calculate the concentration of $H_2O_2$(mol/L) with the volume (mL) of $KMnO_4$ consumed.

## 1.3.2 Determination of reactionvelocity

(1) Take five clean 50 mL conical bottles and label as 1–5. Follow the below Table 12.1 to set up a test system.

Table 12.1 **Test system**

| Reagent(mL) | 1 | 2 | 3 | 4 | 5 |
|---|---|---|---|---|---|
| $H_2O_2$(0.08 mol/L) | 0.50 | 1.00 | 1.50 | 2.00 | 2.50 |
| Distilled water | 3.00 | 2.50 | 2.00 | 1.50 | 1.00 |

(2) Shake well at room temperature. Add 0.5 mL of 1 : 1,000 diluted blood to each bottle one by one and keep shaking during the process, and place in room temperature for 5 min.

(3) Add 2.0 mL of 25% $H_2SO_4$ to each bottle in order (also keep shaking) to stop the enzyme reaction immediately.

(4) Finally, titrate each bottle to reddish color with 0.004 mol/L $KMnO_4$, and record the amount of $KMnO_4$ consumed (mL).

## 1.3.3 Calculation

(1) Calculate the concentration of $H_2O_2$ in each bottle with the following equation:

$$\text{the concentration of } H_2O_2(\text{mol/L}) = \frac{H_2O_2(\text{mol/L}) \times H_2O_2 \text{ added (mL)}}{4(\text{mL})} = \frac{H_2O_2(\text{mmol})}{4(\text{mL})}$$

(In this equation, 4 represents the 4 mL amount of reactive liquid.)

(2) Calculate the reaction velocity according to the consumed amount of $H_2O_2$(mmol) in the reaction:

reaction velocity ($V$) = added amount of $H_2O_2$(mmol) −remained amount of $H_2O_2$(mmol)

i.e., $H_2O_2$(mol/L) × added volume(mL) −$KMnO_4$(0.004 mol/L) ×the volume of $KMnO_4$ consumed× 5/2

(In this equation, 5/2 represents the mol conversion coefficient of the reaction between $KMnO_4$ and $H_2O_2$.)

# 1.4   Results and discussion

## 1.4.1   Results

Using the results of an experiment as an example to show how to calculate the $K_m$ value of catalase for reference.

In here, the concentration of $KMnO_4$ is 0. 004 mol/L, the calibration concentration of $H_2O_2$ is 0.08 mol/L. All calculations are shown in the following Table 12. 2.

Table 12.2   Calculation result

| Calculation procedure | 1 | 2 | 3 | 4 | 5 |
|---|---|---|---|---|---|
| ① Added $H_2O_2$ ( mL) | 0. 50 | 1. 00 | 1. 50 | 2. 00 | 2. 50 |
| ② Added amount of $H_2O_2$ ( mmol) = ①×0. 08 | 0. 04 | 0. 08 | 0. 12 | 0. 16 | 0. 20 |
| ③ Concentration of substrate [ S]  = ②÷4 | 0. 01 | 0. 02 | 0. 03 | 0. 04 | 0. 05 |
| ④ Consumed amount of $KMnO_4$( mL) | 1. 35 | 3. 70 | 6. 40 | 9. 80 | 13. 20 |
| ⑤ Remained $H_2O_2$( mmol) = ④ ×0. 004 × 5/2 | 0. 0135 | 0. 037 | 0. 064 | 0. 098 | 0. 132 |
| ⑥ Reaction velocity( $V$ )  =②−⑤ | 0. 0265 | 0. 043 | 0. 056 | 0. 062 | 0. 068 |
| ⑦ [ S]/$V$=③÷⑥ | 0. 377 | 0. 465 | 0. 536 | 0. 645 | 0. 735 |

According to the result in Table 12. 2, draw a Hanes−Woolf plot by using [ S]/$V$ as $y$−axis and [ S] as $x$−axis. $K_m$ can be determined from the intercept on $x$−axis( $-K_m$)

At last, obtain $K_m$ =0. 032 mol/L.

## 1.4.2   Discussion

( 1) What are the factors that affect the activity of enzymes?

There are many factors that can affect an enzyme's activity. These factors are generally referred to as effectors. Examples of effectors are temperature, pH, inhibitors, and activators. Effectors such as temperature and pH generally affect the noncovalent interactions that occur in an enzyme's structure. Inhibitors and activators can affect the enzyme's structure and/or its interaction with a substrate, prosthetic group, or another ligand.

( 2) What is the purpose of each experimental procedure?

Under all assay conditions, the enzyme will be incubated with the excess substrate ( hydrogen peroxide, $H_2O_2$) for 5 minutes on ice. The reaction will then be quenched ( stopped) by the addition of $H_2SO_4$. The amount of $H_2O_2$ remaining in the reaction mixture ( enzyme + substrate + cofactor +buffer) after 5 minutes of catalase action will be determined by titration with potassium permanganate ( $KMnO_4$) , a very strong oxidizing reagent. The amount of substrate remaining in the mixture is inversely proportional to the activity of the enzyme.

*Wang Huilian*

# Chapter 13

## Isolation, Purification, and Identification of Enzymes

### Experiment 1   Separation and Purification, Specific Activity Determination and Kinetic Analysis of AKP

Alkaline phosphatase (AKP) exists widely in the biological world. It can release inorganic phosphorus by hydrolyzing phosphate bond specifically. The activity of AKP can be used as an important biochemical index for the diagnosis of skeletal and liver diseases. AKP is a biocatalytic macromolecule whose reaction rate is easily affected by environmental factors such as temperature and pH. It is of great significance to master the kinetic properties of the enzyme.

In this experiment, alkaline phosphatase is isolated from pig liver tissue and its enzymatic properties will be studied. Alkaline phosphatase is isolated from liver homogenate by organic solvent precipitation method. That is to say, sodium acetate is used to rupture cell membrane, magnesium acetate is used to stabilize and protect the enzyme, n-butyl alcohol is used to remove some of the impurity protein, and cold acetone is used to precipitate the enzyme. After all these procedures, relatively purified AKP can be obtained by centrifugation. With phenyldisodium phosphate (the reaction substrate) hydrolyzed by the prepared AKP, the phenol will be produced. The phenol will react with potassium ferricyanide under the condition of alkaline solution and 4-aminoantipyrine, which shows red color (benzenedium phosphate method). The enzyme activity and the influence of various factors on the enzyme can be determined by the depth of red color. The separation and purification of AKP, and specific activity determination and kinetic analysis were shown in Figure 13.1.

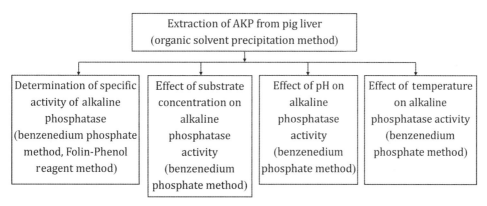

Figure 13.1   AKP separation and purification, specific activity determination and kinetic analysis

# Experiment 1A   Separation and purification, specific activity determination of AKP

## 1.1   Principle

The method of enzyme extraction is similar to that of protein extraction. The alkaline phosphatase is extracted from pig liver homogenate by cold acetone precipitation method. The specific activity of AKP in liver tissue is determined by the number of milligrams of protein per milliliter of the sample [ U/( mg · pr)], based on the specific activity of the enzyme expressed as the enzyme activity per milligram of protein, then the specific activity of alkaline phosphatase in liver tissue is determined.

## 1.2   Materials

### 1.2.1   Reagents and solutions

(1) Sample: fresh pig liver.

(2) 0.5 mol/L magnesium acetate solution: dissolve 10.72 g magnesium acetate in distilled water, qs to 100 mL.

(3) 0.1 mol/L sodium acetate solution: dissolve 0.82 g sodium acetate in distilled water, qs to 100 mL.

(4) 0.01 mol/L magnesium acetate −0.01 mol/L sodium acetate solution: take 2 mL of 0.5 mol/L magnesium acetate solution and 10 mL of 0.1 mol/L sodium acetate solution, mix well, qs to 100 mL with distilled water.

(5) 0.01 mol/L Tris −0.01 mol/L magnesium acetate pH 8.8 buffer: dissolve 1.21 g three hydroxym-ethyl methylamino methane in distilled water to 100 mL to make 0.1 mol/L Tris solution. Take 10 mL of 0.1 mol/L Tris solution, add 80 mL distilled water and 2 mL of 0.5 mol/L magnesium acetate solution, then adjust pH to 8.8 with 1% glacial acetic acid, qs to 100 mL with distilled water.

(6) 0.04 mol/L substrate solution: dissolve 1.05 g disodium phenyl phosphate in distilled water after boiling and cooling, qs to 100 mL. Add 0.4 mL chloroform as antiseptic, and store the solution into a brown

bottle in the refrigerator.

(7) 3% aminoantipyrine: dissolve 0.3 g 4-aminotepyrine and 4.2 g sodium bicarbonate in distilled water, qs to 100 mL. Store the solution into a brown bottle in the refrigerator.

(8) 0.5% potassium ferricyanide: dissolve 0.5 g potassium ferricyanide and 1.5 g boric acid in 40 mL distilled water, respectively. Mix them together, qs to 100 mL with distilled water and stored into a brown bottle in the refrigerator.

(9) 0.1 mg/mL protein standard solution: dissolve 0.1 mg bovine serum albumin with normal saline, qs to 100 mL.

(10) 0.5 mol/L sodium hydroxide solution: dissolve 2 g sodium hydroxide with distilled water, qs to 100 mL.

(11) Alkaline solution: dissolve 2 g sodium hydroxide in distilled water, qs to 100 mL to make 0.5 mol/L sodium hydroxide solution. Dissolve 5.3 g sodium carbonate in distilled water, qs to 100 mL to make 0.5 mol/L sodium carbonate solution. Take 20 mL these two solutions respectively and mix them in distilled water, qs to 100 mL to make an alkaline solution.

(12) Alkaline copper reagent: reagent A: dissolve 0.4 g sodium carbonate, 0.4 g sodium hydroxide, and 0.05 g sodium hydroxide in distilled water, qs to 100 mL. Reagent B: dissolve 1 g copper sulfate pentahydrate in distilled water, qs to 100 mL. Mix reagent A with reagent B in terms of the ratio of 10 : 1 to make alkaline copper reagent.

(13) Phenol reagent: N-butanol, acetone, 95% ethanol (all analytical pure).

## 1.2.2 Special equipment

①Suction pipet; ②cylinder; ③mortar; ④graduated centrifuge tube; ⑤qualitative filter paper; ⑥centrifuge; ⑦glass funnel; ⑧counter balance; ⑨thermostat water bath; ⑩visible range spectrophotometer.

# 1.3 Methods

## 1.3.1 Extraction of alkaline phosphatase

(1) Homogenate preparation

1) Weigh 2 g fresh pig liver tissue and cut into pieces, add 6.0 mL of 0.01 mol/L magnesium acetate-0.01 mol/L sodium acetate solution, grind to the homogenate.

2) Pour the homogenate intoa scale centrifuge tube, get solution A. Record $V_A$.

3) Add 4.9 mL of Tris buffer(pH 8.8) into a clean test tube. Add 0.1 mL solution A, get solution $A_1$ (1 : 50) to measure the specific activity.

(2) Elimination of impurity protein

Add 2 mL n-butanol c to solution A, stir with glass rod fully for 2 min, and place it at room temperature for 20 min. Filter and collect the filtrate.

(3) AKP precipitation with acetone

Add the same volume of cold acetone into the filtrate, and mix immediately. Centrifuge for 5 min at 2,000 r/min, and discard the supernatant.

(4) Dissolution of AKP

1) Add 4.0 mL of 0.5 mol/L magnesium acetate into precipitate, stir it with a glass rod and dissolve it, and get solution B. Record $V_B$.

2) Add 4.9 mL Tris buffer (pH 8.8) into a clean test tube. Take 0.1 mL solution B into the above solution, and get solution $B_1$(1 : 50) to measure the specific activity and will use in the kinetic analysis.

## 1.3.2   Determination of alkaline phosphatase activity

Label 4 dry tubes as shown in Table 13.1.

Table 13.1   Procedure for determination of alkaline phosphatase activity

| Reagent (mL) | Tube | | | |
| --- | --- | --- | --- | --- |
| | Blank tube | Standard tube | Test tube $A_1$ | Test tube $B_1$ |
| pH 8.8 Tris buffer | 1.0 | – | – | – |
| 0.04 mol/L substrate solution | 1.0 | 1.0 | 1.0 | 1.0 |
| 37 ℃ water bath for 5 min | | | | |
| 0.01 mg/mL standard phenol | – | 1.0 | – | – |
| Enzyme solution to be tested | – | – | 1.0 | 1.0 |
| Incubate in 37 ℃ for 15 min | | | | |
| 0.5 mol/L NaOH solution | 1.0 | 1.0 | 1.0 | 1.0 |
| 3% 4-amino-antipyrine | 1.0 | 1.0 | 1.0 | 1.0 |
| 0.5% potassium ferricyanide | 2.0 | 2.0 | 2.0 | 2.0 |

Shake well and stand still at room temperature for 10 min. Measure the absorbance at 510 nm.

## 1.3.3   Determination of protein content

Label 4 dry tubes as shown in Table 13.2.

Table 13.2   procedure for the determination of protein content

| Reagent (mL) | Tube | | | |
| --- | --- | --- | --- | --- |
| | Blank tube | Standard tube | Test tube $A_1$ | Test tube $B_1$ |
| pH 8.8 Tris buffer | 1.0 | – | – | – |
| Solution to be tested | – | – | 1.0 | 1.0 |
| Protein standard solution | – | 1.0 | – | – |
| Alkaline copper reagent | 5.0 | 5.0 | 5.0 | 5.0 |
| Shake well and stand still 10 min at room temperature | | | | |
| Phenolic reagents | 0.5 | 0.5 | 0.5 | 0.5 |

Shake well at room temperature and stand still for 25 min, and measure the absorbance at 650 nm.

## 1.3.4   Notes

(1) Strictly follow the procedure. Transfer the accurate number of reagents.

(2) Prepare ready-to-use reagents before use, store at the refrigerator for no more than one week.

(3) Place the reagents that need to be treated at low temperature in the refrigerator at 4 ℃.

(4) Add organic solvent slowly and stir gently to avoid heating-up and denaturation due to local high concentration.

(5) In the color reaction, the tubes must be fully shaken and then stand still for reaction.

# 1.4 Results and discussion

## 1.4.1 Calculation

(1) Calculate the number of active units of AKP

The number of AKP active units ( U/m) in per milliliter enzyme solution to be tested = $A_{test}/A_{standard} \times C_{standard}$.

(2) Calculate the protein concentration of enzyme solution to be tested

The protein concentration of enzyme solution to be tested ( mg/mL) = $A_{test}/A_{standard} \times C_{standard}$.

(3) Calculate the specific activity of AKP

The specific activity of AKP [ ( U/( mg · pr)] = The number of AKP active units ( U/m) in per milliliter enzyme solution to be tested/Protein concentration in enzyme solution to be tested ( mg/mL) .

## 1.4.2 Original data record

(1) Data record from the determination of the absorbance at 510 nm and 650 nm ( Table 13.3) .

Table 13.3 Data from the determination

| Reagent | Tube | | | |
| --- | --- | --- | --- | --- |
| | Blank tube | Standard tube | Test tube $A_1$ | Test tube $B_1$ |
| The absorbance of AKP activity (510 nm) | | | | |
| The absorbance of protein content determination (650 nm) | | | | |

(2) The correlated data record of AKP specific activity calculation ( Table 13.4) .

Table 13.4 The correlated data of AKP specific activity

| Enzyme solution to be tested | Protein concentration per milliliter | Number of enzyme activity units per milliliter | Specific activity | Purified fold |
| --- | --- | --- | --- | --- |
| $A_1$ | | | | |
| $B_1$ | | | | |

# Experiment 1B    Effect of substrate concentration of on the activity of alkaline phosphatase

## 1.1    Principle

Under the circumstances of other unchanged conditions, the reaction velocity ($v$) of enzymatic increases with the substrate concentration ($[S]$). When the reaction rate reaches a certain rate, it does not increase with the increase of substrate concentration and gradually tends to a certain value, which is called the maximum reaction velocity ($v_{max}$). The corresponding substrate concentration at $1/2\ v_{max}$ is defined as the Michaelis constant ($K_m$, mol/L). The relationship between substrate concentration and reaction rate can be expressed by Michaelis−Constant equation (Equation 13.1). The figure presented by the equation is shown in Figure 13.2. The $v_{max}$ in the graph is only infinitely approaching to a certain value, and the $K_m$ corresponding to $1/2\ v_{max}$ is also infinitely approaching to a certain value. If the reciprocal is conducted on both sides of the Michaelis−Constant, the equation of 13.2 is obtained, and the double reciprocal plot, known as the Lineweaver−Burk plot can be drawn as Figure 13.3. The intercept on the $x$−axis and $y$−axis can be used to determine $K_m$ and $v_{max}$.

$$v = \frac{v_{max}[S]}{K_m + [S]}$$    (Equation 13.1)

$$\frac{1}{v} = \frac{K_m}{v_{max}} \cdot \frac{1}{[S]} + \frac{1}{v_{max}}$$    (Equation 13.2)

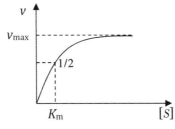

Figure 13.2    Effect of substrate concentration on the rate of enzymatic reaction

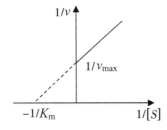

Figure 13.3    Double reciprocal plot

In the experiment, disodium phenyl phosphate used as the substrate is hydrolyzed by the prepared alkaline phosphatase to produce phenol which will react with potassium ferricyanide under the effect of 4−aminoantipyrine and alkaline solution to produce the red compound. The absorbance is determined at 510 nm.

Taking $1/[S]$ as the $x$-axis, $1/v$ as the $y$-axis, the $K_m$ and $v_{max}$ values of alkaline phosphatase could be obtained by extending the straight line, the influence of substrate concentration on the speed of the enzymatic reaction can be analyzed.

# 1.2    Materials

## 1.2.1    Reagents and solutions

(1) 0.04 mol/L substrate solution: dissolve 1.05 g disodium phenyl phosphate in distilled water after boiling and cooling, qs to 100 mL and add 0.4 mL chloroform as antiseptic. Store the solution in a brown bottle in the refrigerator.

(2) 3% 4-aminoantipyrine: dissolve 0.3 g 4-aminotepyrine and 4.2 g sodium bicarbonate in distilled water, qs to 100 mL. Store the solution in a brown bottle and transfer it to the refrigerator.

(3) 0.5% potassium ferricyanide: dissolve 0.5 g potassium ferricyanide and 1.5 g boric acid in 40 mL of distilled water respectively. Mix these two solutions, qs to 100 mL. Store the solution in a brown bottle and transfer it to the refrigerator.

(4) Alkaline solution: dissolve 2 g sodium hydroxide in distilled water to 100 mL to make 0.5 mol/L sodium hydroxide solution; Dissolve 5.3 g sodium carbonate in distilled water to 100 mL to make 0.5 mol/L sodium carbonate solution. Take 20 mL the two solutions respectively, and mix this two solutions, qs to 100 mL to make alkaline solution.

(5) 0.1 mol/L carbonate buffer solution pH 10: dissolve 0.64 g anhydrous sodium bicarbonate and 0.34 g sodium bicarbonate in distilled water, respectively, qs to 100 mL.

(6) 0.10 mg/mL phenol standard solution: dissolve 0.15 g crystalline phenol in 0.1 mol/L hydrochloric acid, qs to 100 mL.

## 1.2.2    Special equipment

①Pipettes; ②test tube; ③constant temperature water bath pot; ④visible spectrophotometer.

# 1.3    Methods

## 1.3.1    Procedure

Label 7 tubes and follow the steps according to Table 13.5.

Table 13.5　Operational steps for the effect of substrate concentration on AKP activity

| Reagent(mL) | Blank | Standard | 1 | 2 | 3 | 4 | 5 |
|---|---|---|---|---|---|---|---|
| Buffer bicarbonate | 0.7 | 0.7 | 0.7 | 0.7 | 0.7 | 0.7 | 0.7 |
| 0.04 mol/L substrate solution | – | – | 0.05 | 0.10 | 0.20 | 0.30 | 0.40 |
| 0.10 mg/mL standard phenol | – | 0.20 | – | – | – | – | – |
| Distilled water | 1.20 | 1.10 | 1.15 | 1.10 | 1.00 | 0.90 | 0.8 |
| 37 ℃ water bath 5 min | | | | | | | |
| Enzyme solution ($B_1$) | 0.10 | – | 0.10 | 0.10 | 0.10 | 0.10 | 0.10 |
| Incubate in 37 ℃ for 15 min | | | | | | | |
| Alkaline solution | 1.0 | 1.0 | 1.0 | 1.0 | 1.0 | 1.0 | 1.0 |
| 3% 4-aminoantipyrine | 1.0 | 1.0 | 1.0 | 1.0 | 1.0 | 1.0 | 1.0 |
| 0.5% potassium ferricyanide | 2.0 | 2.0 | 2.0 | 2.0 | 2.0 | 2.0 | 2.0 |

Shake well and stand still for 10 min, and determine the absorbance at 510 nm.

### 1.3.2　Calculation

The reaction velocity is represented by the amount of phenol product produced at a certain time, and the reaction velocity of each tube is represented by Equation 13.3.

Reaction velocity $v$ (mg phenol/min) = $A_{determining}/A_{standard} \times 0.2 \times 0.1 \times 1/15$ (Equation 13.3)

### 1.3.3　Notes

(1) Follow the procedure strictly. Transfer the accurate number of reagents.
(2) After adding the enzyme solution, incubate for 15 min.
(3) In the color reaction, shake the tubes completely and then stand still for reaction.

## 1.4　Results and discussion

(1) Original data record, substrate concentration and calculation of reaction rate (Table 13.6).

Table 13.6　Original data of substrate concentration and calculation of reaction rate

| Tube | 1 | 2 | 3 | 4 | 5 | Standard |
|---|---|---|---|---|---|---|
| Subsatrte [S] (mol/L) | | | | | | |
| 1/[S] | | | | | | |
| Absorbance (510 nm) | | | | | | |
| Reaction rate (mg phenol/min) | | | | | | |
| 1/v | | | | | | |

(2) Taking [S] as the $x$-axis and $v$ as the $y$-axis, draw the points on the coordinate paper and connect these points in order to observe the shape of the graph.
(3) Taking 1/[S] as the $x$-axis and 1/$v$ as the $y$-axis, draw the points and connect them to a straight

line on the coordinate paper. Obtain the $K_m$ and $v_{max}$ value of alkaline phosphatase by extending the straight line.

# Experiment 1C   Effect of pH on the activity of alkaline phosphatase

## 1.1   Principle

The optimum pH of most enzymes in the human body is about 7.0. The effect of pH on enzyme activity is significant. When the reaction velocity of the enzyme is maximum, the corresponding pH is the optimal pH of the enzyme. If the pH value of the solution is higher or lower than this optimal pH, the enzyme activity will be inhibited and the enzymatic reaction rate is decreased. The enzyme protein will be denatured and even inactivated due to hyperacidity or hyper alkalinity. Under the circumstances of other conditions unchanged, the enzymatic reaction is carried out under a series of different pH conditions. The enzyme activity is determined by the method of disodium phenyl phosphate. Using pH as the $x$-axis and reaction rate as the $y$-axis, the curve of pH effect on enzyme activity can be drawn and the optimum pH range of enzyme is obtained.

## 1.2   Materials

### 1.2.1   Reagent and solutions

(1) Working solution: dissolve 0.6 g disodium phenyl phosphate and 0.3 g 4-aminoantipyrine in distilled water after boiling and cooling, respectively. Mix the two solutions and dilute to 100 mL. Add 0.4 mL chloroform as an antiseptic in the solution and put it into a brown bottle. When using this solution, mix it with the same amount of water.

(2) 0.5% potassium ferricyanide: dissolve 0.5 g potassium ferricyanide and 1.5 g boric acid in 40 mL distilled water, respectively. Mix this two solution and add distilled water to the volume of 100 mL. Store the solution in a brown bottle and transfer it to the refrigerator.

(3) Alkaline solution: dissolve 2 g sodium hydroxide in distilled water, qs to 100 mL to make 0.5 mol/L sodium hydroxide solution. Dissolve 5.3 g sodium carbonate in distilled water, qs to 100 mL to make 0.5 mol/L sodium carbonate solution. Take 20 mL the two solutions respectively, and mix them and dilute to 100 mL to make an alkaline solution.

(4) 0.2 mol/L glycine solution: dissolve 1.50 g glycine in distilled water to 100 mL.

(5) 0.2 mol/L sodium hydroxide solution: dissolve 0.8 g sodium hydroxide in distilled water to 100 mL.

(6) Preparation of different pH buffers (Table 13.7).

Table 13.7   Preparation of different pH buffers

| Reagent( mL) | Tube | | | | | | |
|---|---|---|---|---|---|---|---|
| | 1 | 2 | 3 | 4 | 5 | 6 | 7 |
| 0.2 mol/L glycine solution | 5 | 5 | 5 | 5 | 5 | 5 | 5 |
| 0.2 mol/L sodium hydroxide solution | 0.1 | 0.3 | 0.8 | 1.9 | 3.1 | 4.9 | 5.4 |
| Distilled water | 5.9 | 5.7 | 5.2 | 4.1 | 2.9 | 1.1 | 0.6 |
| pH of the buffer | 8 | 8.5 | 9 | 9.5 | 10 | 11 | 12 |

## 1.2.2   Special equipment

①Pipettes; ②test tube; ③constant temperature water bath pot; ④visible spectrophotometer.

# 1.3   Methods

## 1.3.1   Procedure

Label 8 tubes as 0-7 and follow the steps according to Table 13.8.

Table 13.8   Operation steps of pH effect on AKP activity

| Reagent( mL) | Tube | | | | | | | |
|---|---|---|---|---|---|---|---|---|
| | 0 | 1 | 2 | 3 | 4 | 5 | 6 | 7 |
| Corresponding pH value of each tube | 10 | 8 | 8.5 | 9 | 9.5 | 10 | 11 | 12 |
| Buffer | 0.5 | 0.5 | 0.5 | 0.5 | 0.5 | 0.5 | 0.5 | 0.5 |
| Enzyme solution($B_1$) | – | 0.1 | 0.1 | 0.1 | 0.1 | 0.1 | 0.1 | 0.1 |
| Water bath at 37 ℃ for 5 min | | | | | | | | |
| 37 ℃ pretreatment working solution | 3.0 | 3.0 | 3.0 | 3.0 | 3.0 | 3.0 | 3.0 | 3.0 |
| Mix well, water bath at 37 ℃ for 15 min | | | | | | | | |
| Alkaline solution | 1.0 | 1.0 | 1.0 | 1.0 | 1.0 | 1.0 | 1.0 | 1.0 |
| 0.5% potassium ferricyanide | 2.0 | 2.0 | 2.0 | 2.0 | 2.0 | 2.0 | 2.0 | 2.0 |
| Enzyme solution($B_1$) | 0.1 | – | – | – | – | – | – | – |

Mix well, stand still at room temperature for 10 min, and measure the absorbance at 510 nm.

## 1.3.2   Notes

(1) Follow the procedure strictly. Transfer the accurate number of reagents.
(2) After adding the enzyme and working solution, incubate for 5 min and 15 min, respectively.
(3) In the color reaction, shake the tubes completely and then stand still for reaction.

# 1.4 Results and discussion

The original data should be recorded in Table 13.9.

Table 13.9 Original data record

| | Tube | | | | | | | |
|---|---|---|---|---|---|---|---|---|
| | 0 | 1 | 2 | 3 | 4 | 5 | 6 | 7 |
| pH | 8 | 8.5 | 9 | 9.5 | 10 | 11 | 12 | 10 |
| $A_{510}$ | | | | | | | | |

Taking pH as the $x$-axis and $A_{510}$ as the $y$-axis, plot the curve of the effect of acidity and alkalinity on the activity of alkaline phosphatase to determine the optimum pH range.

# Experiment 1D   Effect of temperature on the activity of alkaline phosphatase

# 1.1 Principle

The optimum temperature of most enzymes in the human body is about 37 ℃. The effect of temperature on enzyme activity is significant and has a dual effect. At low temperature, the enzyme activity is inhibited, and with the increase of temperature, the reaction rate of enzyme accelerates. When the reaction rate reaches the maximum, the temperature is called the optimum temperature of the enzyme. If the temperature continues to rise, the enzyme protein will denature gradually and the enzymatic reaction rate decreases or even deactivates. Under the circumstances of other conditions unchanged, the enzymatic reaction is carried out under a series of different temperature conditions and the activity of the enzyme is determined by the method of disodium phenyl phosphate. The temperature is used as the $x$-axis and the reaction rate as the $y$-axis, and the curve of the effect of temperature on the enzyme activity can be drawn. Thus, the optimum temperature will be obtained.

# 1.2 Materials

## 1.2.1 Reagent and solutibn

(1) Working solution: Experiment 1C, Chapter 13.
(2) 0.5% potassium ferricyanide: Experiment 1C, Chapter 13.
(3) Alkaline solution: Experiment 1C, Chapter 13.
(4) 0.1 mol/L carbonate buffer solution pH 10: dissolve 0.64 g anhydrous sodium bicarbonate and 0.34 g sodium bicarbonate in distilled water, qs to 100 mL.
(5) 0.10 mg/mL phenol standard solution: dissolve 0.15 g crystalline phenol in 0.1 mol/L hydrochlo-

ric acid, qs to 100 mL.

## 1.2.2   Special equipment

①Pipettes; ②test tube; ③constant temperature water bath pot; ④visible spectrophotometer.

# 1.3   Methods

## 1.3.1   Procedure

Label 8 tubes as 0–7 and follow the steps Table 13.10.

Table 13.10   Operation steps of temperature effect on AKP activity

| Reagent( mL) | Tube | | | | | | | |
|---|---|---|---|---|---|---|---|---|
| | 0 | 1 | 2 | 3 | 4 | 5 | 6 | 7 |
| Reaction temperature | 37 | Room T | 30 | 35 | 37 | 40 | 60 | 80 |
| Enzyme solution ($B_1$) | – | 0.1 | 0.1 | 0.1 | 0.1 | 0.1 | 0.1 | 0.1 |
| 37 ℃ pretreatment working solution | 3.0 | 3.0 | 3.0 | 3.0 | 3.0 | 3.0 | 3.0 | 3.0 |
| Water bath at different temperatures, and incubate for 15 min | | | | | | | | |
| Alkaline solution | 1.0 | 1.0 | 1.0 | 1.0 | 1.0 | 1.0 | 1.0 | 1.0 |
| 0.5% potassium ferricyanide | 2.0 | 2.0 | 2.0 | 2.0 | 2.0 | 2.0 | 2.0 | 2.0 |
| Enzyme solution | 0.1 | – | – | – | – | – | – | – |

Mix well, stand still at room temperature for 10 min, and measure the absorbance at 510 nm.

## 1.3.2   Notes

(1) Follow the procedure strictly. Transfer the accurate number of reagents.
(2) After addingthe enzyme solution, incubate for 15 min.
(3) In the color reaction, shake the tubes completely and then stand still for reaction.

# 1.4   Results and discussion

The original data should be recorded in Table 13.11.

Table 13.11   Original data record

| | Tube number | | | | | | | |
|---|---|---|---|---|---|---|---|---|
| | 0 | 1 | 2 | 3 | 4 | 5 | 6 | 7 |
| Reaction temperature (℃) | 37 | Room T | 30 | 35 | 37 | 40 | 60 | 80 |
| $A_{510}$ | | | | | | | | |

Taking the temperature as the $x$–axis and $A_{510}$ as the $y$–axis, plot the influence curve of temperature on the activity of alkaline phosphatase to determine the optimal temperature.

# Experiment 2   Isolation, Purification and Identification of $\alpha_1$ –antitrypsin ($\alpha_1$ –AT) from Human Serum

Alpha–1–antitrypsin or $\alpha_1$ –antitrypsin ($\alpha_1$ –AT) is a glycoprotein of 52 kDa and is present in approximately equal concentrations in plasma and interstitial fluid. As a protease inhibitor, it is also known as alpha1–proteinase inhibitor ($\alpha_1$ –P1) or alpha1–antiproteinase ($\alpha_1$ –AP) because it inhibits various proteases (not just trypsin). Also referred to as serine protease inhibitor ($\alpha_1$ –proteinase inhibitor or $\alpha_1$ –PI), it inhibits the serine protease trypsin, chymotrypsin as well as pancreatic and especially granulocytic elastase.

The increased levels of $\alpha_1$ –AT are common as it is an acute phase reactant. The elevated levels are also seen in the late pregnancy and during oestrogen therapy because the synthesis of $\alpha_1$ –antitrypsin is stimulated by oestrogens. The low levels of $\alpha_1$ –antitrypsin, on the other hand, are found in neonatal respiratory distress syndrome, severe neonatal hepatitis, preterminal disease of the pancreas and in severe protein–losing enteropathies. However, the measured level of $\alpha_1$ –AT does not drop below the normal range for these $\alpha_1$ –AT–deficiency cases mentioned above. It is slightly confusing.

Here we use $\alpha_1$ –AT as an example to learn all the methods involved in the protein isolation, purification and identification for the following reasons. Firstly, $\alpha_1$ –AT has been well characterized. Secondly, all the procedures are suitable for current teaching practice and can be applied to your future research equally.

This section will conclude the next several experiments. The salting out method will be performed to make primary isolation of proteins from the crude samples, and gel filtration chromatography will be used to remove the salts from the protein samples isolated in the previous salting–out experiment. DEAE–cellulose ion exchange chromatography is applied for further purification, and then the purified protein will be quantified and its enzymatic activity will be determined. The SDS–polyacrylamide gel electrophoresis (SDS–PAGE) will be used to identify the protein on the molecular size basis.

# Experiment 2A   Protein precipitation and protein fraction

## 2.1   Principle

The solubility of proteins depends on, among other things, the salt concentration in solution. At low concentration, salt stabilizes the charged groups on a protein molecule, thus enhancing the solubility of proteins in solutions. This is commonly known as salting–in. As the salt concentration increases, the protein solubility increased accordingly (in most cases) and then reaches the maximum point. Higher salt concentration means that less water is available to solubilize proteins. Therefore, as the salt concentration continues to increase, protein molecules start to precipitate since there are no enough water molecules to interact with proteins. This phenomenon of protein precipitation in the presence of excess salt is known as salting–out.

In this experiment, either a saturated salt solution or powdered salt crystals are slowly added to the protein solution to bring the salt concentration up gradually.

For example, the salt concentration is 25% saturated when 3.0 mL of the salt–free protein solution is mixed with 1.0 mL of the saturated salt solution; 50% when mixed with 3.0 mL of saturated solution and 75% when mixed with 9 mL of saturated solution.

Since different proteins may differ in their charged groups, they could be precipitated at different salt concentrations. The partial collection of the precipitated proteins at different salt concentrations is referred to as fractional precipitation.

## 2.2   Materials

### 2.2.1   Reagents and solutions

(1) Human serum, 15 mL.
(2) Saturated $(NH_4)_2SO_4$ solution, 15 mL.
(3) $(NH4)_2SO_4$ crystals.
(4) 0.05 mol/L, pH 6.4 phosphate buffer solution (PBS).

### 2.2.2   Special equipment

①Graduate cylinder; ②balance; ③centrifuge; ④centrifuge tube; ⑤filtration devices (filter pump, filter flask, funnel, rubber stopper); ⑥beakers; ⑦glass rod; ⑧refrigerator.

## 2.3   Methods

### 2.3.1   First fractionation

(1) Measure 15 mL of human serum and 15 mL of saturated $(NH_4)_2SO_4$ solution with graduated cylinders, respectively.
(2) Pour the latter into the former. Mix them together in a beaker thoroughly with a glass rod.
(3) At this moment, the $(NH_4)_2SO_4$ concentration reaches 50% saturation. Proteins of large molecular weight in human serum may be precipitated at this salt concentration, such as $\alpha_2$-globulin, $\beta$-globulin, $\gamma$-globulin. $\alpha_1$-AT is a kind of $\alpha_1$-globulin.
(4) Store the solution at 4 ℃ for 20 minutes.

### 2.3.2   Centrifugation

(1) Transfer the solution to centrifuge tubes.
(2) Balance the centrifuge tubes by pair.
(3) Centrifuge at 3,000 r/min (rounds per minute) for 20 min.

### 2.3.3   Second fractionation

(1) Take out the centrifuge tubes from the centrifugal rotor gently when it stops completely.
Two layers in the tube can be seen. The top layer is a transparent solution, and the bottom is the protein precipitates.
*Note: never shake the tubes and disturb the flurry bottom layer. $\alpha_1$-AT is in the solution.*
(2) Pour the top layer into a cylinder carefully and measure the volume.
(3) Add an adequate amount of $(NH_4)_2SO_4$ crystals into the solution in the ratio of 17.6 g of $(NH_4)_2SO_4$ crystals per 100 mL solution.
(4) Stir up to dissolve $(NH_4)_2SO_4$ crystals completely.
(5) Store the solution at 4 ℃ for 20 min.

*Note: to accelerate the crystal dissolution, you can use the glass rod to crash the crystals into small pieces. The* $(NH_4)_2SO_4$ *concentration now can be as high as 75%.* $\alpha_1$*-AT, along with some other kinds of proteins in human serum, will be precipitated.*

## 2.3.4　Filtration

(1) Cut two pieces of round filter paper to match the inner diameter of the funnel. The size of the filter paper should be equal to or smaller than the inner diameter of the funnel.

(2) Place these two pieces of filter paper in the funnel. Place the rubber stopper of the funnel on the mouth of the filter flask tightly.

(3) Connect the filter flask to the filter pump with a rubber tube.

(4) Rinse the filter papers with solution, which makes the filter paper wet to stick on the funnel.

(5) Pour the solution into the funnel slowly and then turn on the filter pump. The liquid in the solution will be sucked into the flask and leave the precipitate on the filter paper.

(6) Continue to suck the liquid out until the precipitate on the filter paper becomes paste–like material. Turn off the filter pump.

(7) Transfer the precipitate carefully to a beaker with a spatula.

(8) Add 2 mL of 0.05 mol pH 6.4 PBS in the beaker to dissolve the precipitate. This solution contains a large quantity of $\alpha_1$–AT isolated from human serum.

*Note: when performing this experiment, the glass rod, beakers, and all the glassware should be kept dry to ensure the complement of this experiment.*

# Experiment 2B　Gel filtration chromatography (desalting)

## 2.1　Principle

The technique of gel filtration (exclusion chromatography, gel chromatography, molecular sieve chromatography) was developed in the 1960's.

Gel filtration chromatography separates different proteins, peptides, and even oligonucleotides on the basis of the sizes of the molecules to be separated. The gel is composed of porous beads. When molecules move through a column of porous beads, they can diffuse into the beads to some certain degrees. Smaller molecules diffuse into the pores of the beads deeper and, therefore, move through the column slowly, whereas larger molecules enter less or not at all and then flow through the column more quickly. This is the molecular sieve effect.

As a result, larger and smaller molecules move through the column in different retention time. Both molecular weight and three–dimensional shape affect the degree retention time.

Gel filtration chromatography can be used for analyzing the molecular size, separating components in a mixture, or removing salts from macro biomolecule precipitates.

# 2.2   Materials

## 2.2.1   Reagents and solutions

(1) Sephadex® G-25.

(2) Nessler reagent.

(3) 0.05 mol/L, pH 6.4 phosphate buffer solution (PBS).

(4) Biuret reagent.

## 2.2.2   Special equipment

Column, 20 mm×550 mm.

# 2.3   Methods

## 2.3.1   Set up the column

(1) Clamp the column tube vertically.

(2) Keep the outlet of the column tube closed.

(3) Add PBS (0.05 mol/L, pH 6.4) to the 1/3 volume of the column tube.

(4) Place a plastic beaker below the outlet of the column.

## 2.3.2   Pack the column

(1) Stir up the pre-prepared gel solution in a beaker very carefully using a glass rod. Mix the gel solution gently.

*Note: clean the rod before use or it will create carry-over contamination.*

(2) Add the gel solution into the column to about 2/3 volume of the column. When a clear layer about 1 cm high appears on the top of the opaque gel solution (or 1 cm high of gel sediment forms at the bottom of the gel column), open the valve of the outlet and drain the solution slowly.

(3) Control the flow speed at 30 drops/min until the height of the gel column reaches the 1/3-1/2 of the column height.

*Note: ①When packing the column, keep the stable flow speed to ensure to get a stable and uniform column. ②Leave the PBS solution 2-3 cm above the gel column to avoid being dried due to evaporation. ③The height of the gel is the height of the column after the gel falls down completely.*

## 2.3.3   Balance the column

(1) Connect the outlet of a flask to the column in a rubber tube. Pour PBS (0.05 mol/L, pH 6.4) into the flask.

(2) Adjust the clamp on the rubber tube to balance the flow-in speed of PBS into the column and flow-out speed at the outlet of the column. Measure the pH of the flow-out solution.

(3) Close the outlet when the pH at the inlet and outlet of the column are equal.

*Note: the gel material is very expensive, so we reuse it. The gel in this experiment might be used previously and could be contaminated by ammonia sulfate. The contaminated gel can influence purification quality. Therefore, you need to use Nessler reagent to determine the presence of ammonium ions before adding the sample.*

## 2.3.4   Load samples

(1) Suck out the excess volume of the PBS solution on the top of the gel column using a Pasteur pipette (or dropper), and immediately add the sample solution onto the gel column slowly by draining the sample solution along the inner wall of the column tube.

(2) Open the outlet and control the flow speed at 30 drops/min. When the sample solution enters the gel column completely, wash the inner wall of the column tube with a small amount of PBS.

*Note: adding the sample solution is time-consuming. Be patient. Do not add the sample solution directly onto the top of the gel column to avoid disturbing the flat surface of the column.*

## 2.3.5   Collect the eluted samples

(1) Add 2-3 mL of PBS onto the top of the gel column.

(2) Connect the flask to the gel column using a rubber tube and begin to wash the gel column.

(3) As PBS flows, you can observe a yellow band moving downward on the gel column, which is the protein sample.

(4) When the yellow band reached the bottom, take a small amount of the eluate to react with Biuret reagent.

(5) If the color of the reagent turns to purple, it indicates the proteins are present in this eluate. Begin to collect the eluted solution.

(6) Continue to collect the eluate until the Biuret reagent does not show the purple color.

## 2.3.6   Wash the gel column

(1) Wash the column with PBS at full speed.

(2) Test the collected portion with Nessler reagent. This reagent will turn to orange color once reacting with ammonia sulfate present in the gel beads.

(3) Wash the gel column until the reaction color disappears.

## 2.3.7   Regeneration

(1) Turn the column tube upside down.

(2) Put the mouth of a pipette bulb to the small exit end of the column and squeeze the bulb to flush the beads out to the beaker.

## 2.3.8   Measure

(1) Measure and record the total volume of the eluted sample.

(2) Take 1.0 mL into a small tube for future analysis.

(3) Transfer the remaining into a big plastic test tube, and keep them in the refrigerator.

# Experiment 2C   DEAE-cellulose Ion-exchange chromatography

# 2.1   Principle

Ion-exchange chromatography separates molecules based on the differences in the overall charge of proteins. It is commonly used for protein purification as well as for purification of oligonucleotides, peptides, or other charged molecules.

The protein to be purified must have charges opposite to that of the functional groups on the resin. For example, immunoglobulins, which generally have an overall positive charge, will bind well to cation exchangers, which contain negatively charged functional groups.

Ion-exchange containing diethylaminoethyl (DEAE) or carboxymethyl (CM) groups are most commonly used for this purpose. For example, DEAE is of positive charges, so it can absorb proteins that are negatively charged.

Since this interaction is ionic, the binding must take place under low ionic conditions. Desorption is then brought out by increasing the salt concentration or by altering the pH of the mobile phase.

The ionic properties of DEAE and CM are dependent on pH. Both work well as ion exchangers in the pH range of 4–8 where most protein separations take place. The property of a protein which governs its adsorption to an ion exchanger is the net surface charge. Since the surface charge is the net result of weakly acidic and basic groups of a protein, separation is highly pH dependent. Going from low to high pH, the surface of protein shifts from positively charged to negatively charge. The pH vs. net surface curve is an individual property of a protein and constitutes the basis for selectivity in ion exchange chromatography. At a pH value below its isoelectric point, a protein (that is, positive charged) will adsorb to a cation exchanger such as one containing CM groups. Above the isoelectric point protein (that is, negatively charged) will adsorb to an anion exchanger such as one containing DEAE groups.

Desorption can be brought out by varying pH or ionic strength. When pH value changed, the adsorbed protein has passedits isoelectric point and the surface charges altered too. Thus, varying the pH of the mobile phase can separate and collect the target protein.

At low ionic strengths, all components having binding affinity for the ion exchanger will be tightly adsorbed at the top of the ion exchanger and nothing will be in the mobile phase. Then the ionic strength of the mobile phase will be gradually increased by adding a neutral salt, the salt ions will compete with the protein, and more adsorbed proteins will be partially desorbed and start moving down the column. The more net charges a protein has, the higher the ionic strength is needed to bring about desorption. In order to optimize protein separation in ion exchange chromatography, pH value is carefully chosen that creates sufficiently large net charge differences among the different proteins components.

## 2.2　Materials

### 2.2.1　Reagents and solutions

(1) DEAE-cellulose.
(2) 0.05 mol/L, pH 6.4 phosphate buffer solution.
(3) 0.12 mol/L, pH 6.4 phosphate buffer solution.
(4) Biuret reagent.

### 2.2.2　Special equipment

Column, 10 mm×150 mm.

# 2.3 Methods

## 2.3.1 Set up and pack the column

(1) Follow the procedures described in the section of Gel Filtration Chromatography. The minor difference is that the height of the gel column for ion exchange chromatography is 1/3 of the column with 0.05 mol/L pH 6.4 PBS.

(2) Connect the exit hole of a flask to the column in a rubber tube to keep the flow speed of PBS into the column and out of the column at the exit end.

(3) Place a plastic beaker below the outlet of the column.

(4) Measure the pH of the flow-out solution. Close the exit end when the pH at the entrance and the exit are equal.

*Note: the equilibrium step is very important due to the principle of the DEAE-cellulose ion-exchange chromatography. Because pH can influence adsorption and desorption of proteins and the DEAE-cellulose gel is alkali, the equilibrium time will be slightly longer.*

## 2.3.2 Load sample

Same as that in the section of Gel Filtration Chromatography.

## 2.3.3 Collect the eluted sample

(1) Open the exit end of the column tube and control the flow speed at 15 drop/min.

*Note: control the speed to allow the protein adsorb to the DEAE-cellulose enough.*

(2) Set up a reaction plate for color comparison.

(3) Periodically transfer one drop of the eluate to the hole of the pre-prepared reaction plate.

If the color of the reagent turns purple, it indicates that protein is present in this eluate. But it is the flow-through proteins, not the $\alpha_1$-AT component of our interest since it does not bind to the ion exchanging column. Elute continuously until no more protein can be detected.

*Note: do not discard this portion. Due to many personal reasons, the proteins of interest could be present in this portion. If you can't collect the proteins of interest in the next step, this portion will be used in the later step.*

(4) Use the elution buffer of 0.12 mol/L pH 6.4 PBS to wash out the proteins adsorbed on the column at full speed.

(5) Use the Biuret reaction to determine the collecting point.

When the color of the reagent turns purple, immediately collect the eluted solution.

*Note: this collection should be as quickly as possible since the proteins adsorbed on the lower portion of the column could be eluted out very fast.*

(6) Collect all the eluate until the reaction does not show any color change.

(7) Measure the eluate volume with a cylinder.

(8) Transfer it into the plastic centrifugal tubes.

(9) Mark the tubes and keep them in the refrigerator.

# Experiment 2D   Determination of $\alpha_1$ –AT activity

## 2.1   Principle

The activity of $\alpha_1$ –AT is determined by measuring its capacity of inhibiting the enzymatic activity or trypsin.

Trypsin is an enzyme hydrolyzing peptides. One of its substrates is N−benzoyl−L−arginine−p−nitroanilide ( BAPNA) , a simple peptide. Enzyme trypsin will cleave BAPNA to liberate p−nitroaniline.

p−Nitroaniline is of yellow color whose spectrum is different from BAPNA. The formation of the hydrolytic product, therefore, can be conveniently identified using a simple spectroscopic measurement. $\alpha_1$ –AT has been known as an inhibitor of trypsin to inhibit the trypsin activity partially or completely. Therefore, the activity of $\alpha_1$ –AT can be determined indirectly by comparing the trypsin activity in the presence or absence of $\alpha_1$ –AT.

## 2.2   Materials

### 2.2.1   Reagents and solutions

( 1) BAPNA: dissolve 18 mg BAPNA in 10 mL methanol, add Tris−HCl buffer to 50 mL.
( 2) Trypsin: dissolve 20 mg trypsin in 166 mL pH 3 water with a concentration of 0.12 mg/mL.
(3) 33% HAC.
(4) 0.1 mol/L Tris−HCl buffer.
(5) Samples: serum, the eluate solution of $\alpha_1$ –AT from the experiments above of G−25, DEAE.

### 2.2.2   Special equipment

①722−spectrometer; ②incubator; ③cylinder; ④transporter.

## 2.3   Methods

(1) Label 5 tubes as 1−5. Add the following solutions in each tube according to Table 13.12.

Table 13.12   Reaction system

| Solution | # 1 blank | # 2 control | # 3 serum | # 4 G−25 | # 5 DEAE |
| --- | --- | --- | --- | --- | --- |
| Tris buffer | 1.0 | 0.5 | 0.4 | 0.4 | 0.4 |
| Serum | – | – | 0.1 | – | – |
| G−25 sample | – | – | – | 0.1 | – |
| DEAE sample | – | – | – | – | 0.1 |
| Trypsin | – | 0.5 | 0.5 | 0.5 | 0.5 |

(2) Use 0.1 mol/L Tris-HCl buffer, pH 8.0 containing 0.01 mol/L $CaCl_2$ to adjust the final volume of each reaction mixture to 1.0 mL.

(3) Incubate the test tube at 37 ℃ for 5 min.

(4) Add 2.5 mL freshly prepared BAPNA to each tube.

(5) Incubate at 37 ℃ for 10 min.

(6) Stop the reaction by adding 0.5 mL of acetic acid (33% $v/v$) in each tube.

(7) Measure the $A_{410}$ of each reaction mixture.

(8) Calculate the activity of each mixture. The full enzymatic activity is determined based on the readings of the # 2 tube in which no inhibitor is added in the reaction mixture. The # 1 tube is used for the blank reading on a spectrophotometer.

$$\alpha_1-AT\ activity(mg/mL) = [(A_{control}-A_{test})/A_{control}] \times 0.06 \times (1/0.1)$$

# Experiment 2E　Protein quantitation

## 2.1　Principle

Under alkaline conditions, substances containing two or more peptide bonds can form a purple complex with copper salts in the reagent.

## 2.2　Materials

### 2.2.1　Reagents and solutions

(1) Biuret reagent.

(2) 0.9% NaCl.

(3) Protein standard solution: 7%.

(4) Samples: serum, the eluate solution of $\alpha_1-AT$ from the experiments above of G-25, DEAE.

### 2.2.2　Special equipment

722-spectrometer.

## 2.3　Methods

(1) Label 5 tubes as 1-5. Add the following solutions in each tube according to Table 13.13.

Table 13. 13   **Reaction system**

| Solution | #1 blank | #2 standard | #3 serum | #4 G–25 | #5 DEAE |
|---|---|---|---|---|---|
| 0.9% NaCl | 0. 1 | – | – | – | – |
| Protein standard solution | | 0. 1 | | | |
| Serum | – | – | 0. 1 | – | – |
| G–25 sample | – | – | – | 0. 1 | – |
| DEAE sample | – | – | – | – | 0. 1 |
| Biuret reagent | 5. 0 | 5. 0 | 5. 0 | 5. 0 | 5. 0 |

(2) Mix well and incubate at room temperature for 30 min.

(3) Measure the $A_{520}$. Calculate the concentration of the samples using the formulation below.

$$C_u = A_u / A_s \times C_s$$

# Experiment 2F   Protein separation by SDS–PAGE

Follow the same principles and related procedures as Experiment 2, Chapter 8, Part Ⅱ.

*Wu Ning, Liu Baoqin*

**Part IV**

# Carbohydrate and Lipid Experiments

# Chapter 14

## Determination of Glucose

There are many methods of blood glucose determination, so far, it can be divided into four categories: redox method, shrinkage method, enzymatic method and other methods like gas−liquid chromatography, mass spectrometry, enzyme electrode method, solid phase enzyme multilayer thin film method and so on. Each method has its own advantages and disadvantages, which should be selected according to the specific situation. At present, glucose oxidase method is often used in clinical practice, and hexokinase is a widely recognized reference method for blood glucose determination in the world.

## Experiment 1  Determination of Blood Glucose by Using Glucose Oxidase Method

### 1.1  Principle

Glucose is oxidized by glucose oxidase (GOD) to produce gluconic acid and hydrogen peroxide ($H_2O_2$). Hydrogen peroxide is then oxidatively coupled with 4−amino−antipyrine (4−AAP) and phenol in the presence of peroxidase (POD) to form quinone, that is the Trinder reaction. The quinone is red−violet in color, with the intensity of the color (the absorbance at 505 nm) being in proportion to the glucose concentration. The reaction formula is as follows.

This method is specific to glucose and is not disturbed by other sugars and reductive substances.

$$C_6H_{12}O_6 + O_2 + H_2O \xrightarrow{\text{glucose oxidase}} C_6H_{12}O_7 + H_2O_2$$
$$\text{Glucose} \qquad\qquad\qquad \text{Gluconic Acid}$$

4−aminoantipyrine
(4−AAP)

quinone(red)

# 1.2 Materials

## 1.2.1 Reagents and solutions

(1) Sample: fresh serum.

(2) Standard glucose storage solution(100 mmol/L): weigh 1.802 g glucosum anhydricum (dry in the oven at 80 ℃ in advance and store in desiccators), dissolve it with 12 mmol/L benzoic acid solution and transfer into a 100 mL volumetric bottle, dilute to 100 mL with 12 mmol/L benzoic acid. Place at least 2 h before use. Store in the refrigerator at 4 ℃ for long-stem storage.

(3) Standard glucose standard working solution(5 mmol/L): take 5.0 mL glucose standard storage solution, dilute it to 100 mL with 12 mmol/L benzoic acid solution in a bottle, mix well.

(4) 0.1 mol/L phosphate buffer(pH 7.0): dissolve 8.67 g anhydrous disodium hydrogen phosphate and 5.3 g anhydrous potassium dihydrogen phosphate in 800 mL distilled water, adjust pH to 7.0 with 1 mol/L sodium hydroxide (or 1 mol/L hydrochloric acid), qs to 1 L with distilled water.

(5) Enzyme reagent: dissolve 1,200 U peroxidase, 1,200 U glucose oxidase, 10 mg 4-amino antipyrine, 100 mg sodium azide in 80 mL phosphate buffer, adjust pH to 7.0 with 1 mol/L NaOH, qs to 100 mL with phosphate buffer. Store in the refrigerator at 4 ℃ for three months.

(6) Phenol solution: dissolve 100 mg redistilled phenol in 100 mL distilled water. Store in a brown-color bottle.

(7) Enzyme phenol mix(working solutuion): mix isometric enzyme reagent and phenol solution. Store at 4 ℃ for one month.

(8) 12 mmol/L benzoic acid solution: add 1.465 g benzoic acid to 800 mL distilled water, heating to accelerate dissolution, qs to 1 L with distilled water after cooling down.

## 1.2.2 Special equipment

①Medium test tube (×3); ②pipette 5 mL (×1); ③sample injector; ④water-bath; ⑤spectrophotometer.

# 1.3 Methods

## 1.3.1 Procedure

(1) Reaction system setting

Label 3 tubes, add reagents according to the Table 14.1.

Table 14.1　Reaction system setting

| Reagents( mL) | Blank tube | Standardt ube | Sample tube |
| --- | --- | --- | --- |
| Standard glucose working solution | — | 0.02 | — |
| Serum | — | — | 0.02 |
| Distilled water | 0.02 | — | — |
| Enzyme phenol mix | 3.0 | 3.0 | 3.0 |

(2) Reaction and determination

Mix each tube, and incubate in water bath at 37 ℃ for 15 min.

Zero the spectrophotometer at 505 nm with the blank tube and measure $A_{505}$ of each tube for three times, and calculate the average value.

### 1.3.2　Notes

(1) Use non-hemolytic specimen to prevent intraglobular glucose-6-phosphate enters the serum during hemolysis. Separate serum within 30 minutes. Otherwise, glycolysis will cause the result reduced. The use of potassium oxalate-sodium fluoride as anticoagulant can inhibit the decomposition of glucose. The separated serum ( plasma) specimens can stabilize for 24 h at 2-8 ℃ and for 30 d at -20 ℃.

(2) Glucose oxidase is highly specific to β-D-glucose, about 36% of the glucose in the solution is the α type and 64% is the β type. The complete oxidation of glucose requires mutarotation reaction of α and β type. Some commercial glucose oxidase kits contain glucose mutarotase, which can accelerate this reaction. However, in the end-point method, the process of spontaneous mutarotation can be achieved by prolonging the incubation time. The freshly prepared glucose standard solution is mainly α type, so it is necessary to place more than 2 h ( preferably overnight) to be used after the balance of mutarotation.

(3) The reagents scale in this method is in a small amount, in order to ensure the results are reliable, please add the specimen directly to the reagents and rinse the pipe repeatedly with reagents in operation.

(4) Serum with severe jaundice, hemolytic should be prepared nonprotein filtrate first before measure.

## 1.4　Results and discussion

### 1.4.1　Result

(1) Data record

Record the absorbance of each tube at 505 nm wavelengths, obtain the average of $A_{505}$ from duplicate or triplicate.

(2) Result calculation

$$\text{serum glucose( mmol/L)} = \frac{A_U}{A_S} \times 5$$

$A_U$: $A_{505}$ of the sample tube. $A_S$: $A_{505}$ of the standard tube. **5**: concentration of standard glucose working solution(5 mmol/L)

(3) Normal value

Fasting serum glucose: 3.89-6.11 mmol/L(70-110 mg/dL)

### 1.4.2　Discussion

What factors could influence the determination of blood glucose content using this method, and what are the advantages and disadvantages compared with other methods?

# Experiment 2　Determination of Blood Glucose by Using Hexokinase Method

## 2.1　Principle

Hexokinase(HK) catalyzes the phosphorylation of glucose by adenosine triphosphate (ATP) producing glucose−6−phosphate(G−6−P) and adenosine diphosphate(ADP). Glucose−6−phosphate is dehydrogenated by glucose−6−phosphate dehydrogenase(G−6−PD) to 6−phosphogluconic acid (6−PGA) with the reduction of $NADP^+$ to NADPH and $H^+$. The increase in reductive NADPH production is directly proportional to the glucose concentration, and the concentration of glucose in the blood can be calculated by measure the increase rate of NADPH absorbance, compared with the standard tube at 340 nm wavelength or by the end-point method. The reaction formula is as follows:

$$Glucose+ATP \xrightarrow{HK} glucose\ 6-phosphate+ADP$$
$$Glucose\ 6-phosphate+NADP^+ \xrightarrow{G-6-PD} 6-phosphogluconic\ acid+NADPH+H^+$$

## 2.2　Materials

### 2.2.1　Reagents and solutions

(1) Sample: fresh serum.

(2) Enzyme mix: hexokinase method for blood glucose determination is mostly applied with a commercial kit. At present, there are many domestic and overseas manufacturers for this kit, and the formula is similar. The basic components are as follows(Table 14.2).

Table 14.2　The basic components of enzyme mix

| Components | Concentration |
| --- | --- |
| Triethanolamine salt buffer(pH 7.5) | 50 mmol/L |
| $MgSO_4$ | 2 mmol/L |
| ATP | 2 mmol/L |
| $NADP^+$ | 2 mmol/L |
| HK | ≥1,500 U/L |
| G−6−PD | 2,500 U/L |

Store in refrigerator at 4 ℃, prepare according to the instruction.

(3) 100 mmol/L standard glucose storage solution: see experiment 1, Chapter 14.

(4) 5 mmol/L standard glucose working solution: see experiment 1, Chapter 14.

## 2.2.2　Special equipment

①Medium test tube (×5); ②water-bath; ③ultraviolet spectrophotometer.

# 2.3　Methods

## 2.3.1　Procedure

(1) Reaction system setting

Label 4 test tubes, add reagents according to the Table 14.3.

Table 14.3　Reaction system setting

| Reagents(mL) | Blank tube | Standard tube | Control tube | Sample tube |
| --- | --- | --- | --- | --- |
| Physiological saline solution | 0.02 | — | 2.0 | — |
| Standard glucose working solution | — | 0.02 | — | — |
| Serum | — | — | 0.02 | 0.02 |
| Enzyme mix | 2.0 | 2.0 | — | 2.0 |

(2) Reaction

Mix each tube, and incubate in water bath at 37 ℃ for 10 min.

(3) Colorimetric assay

Zero the spectrophotometer at 340 nm with distilled water, and measure $A_{340}$ of each tube for three times, and calculate the average value.

## 2.3.2　Notes

(1) The condition of the enzymatic reaction should be strictly controlled, such as reaction temperature, time, pH of the buffer, and the accuracy of sample addition in the determination of blood glucose concentration by an enzymatic reaction.

(2) HK is the key enzyme for the reaction and high purity products must be used, its specific activity should be more than 140 U/mg enzyme protein (25 ℃). The heteroenzyme of HK, such as glutathione reductase, glucose-6-phosphate dehydrogenase, should be less than 0.01% HK active unit, and glucose phosphate isomerase should be less than 0.02% HK active unit.

(3) Optimum pH of HK is 6.0-9.0 and pI is 4.5-4.8. $Mg^{2+}$ is the activator of HK and EDTA is an inhibitor. The optimum pH of glucose-6-phosphate dehydrogenase is 8.5 with $NADP^+$ as a coenzyme and 7.8 with $NAD^+$ as a coenzyme.

(4) The purity requirement of $NADP^+$ or $NAD^+$ is over 98%.

(5) Requirements for reagents: the absorbance of reagent blank is less than 0.1 at heat preservation 30 min and less than 0.01 at heat preservation between 6 min and 30 min. The absorbance of 33.3 mol/L glucose standard solution is between 1.6 and 1.8 at heat preservation 30 min and less than 0.01 at heat preservation between 8 min and 30 min.

(6) If a heat-stable glucokinase is used instead of HK, the specificity and the stability can be improved.

(7) Mild hemolysis, lipidemia, jaundice, vitamin C, sodium fluoride, heparin, EDTA and oxalate and so on, do not interfere with the determination of this method, except severe hemolysis because of the organo-

phosphate released from the red blood cells and some enzymes that consume $NADP^+$.

## 2.4 Results and discussion

### 2.4.1 Result

(1) Data record

Record the absorbance of each tube at 340 nm wavelengths in the Table 14.4.

Table 14.4 Data of absorbance ($A_{340}$)

| Number of measurement | Absorbance $A_{340}$ | | | |
| --- | --- | --- | --- | --- |
| | Blank tube $A_B$ | Standard tube $As$ | Control tube $Ac$ | Sample tube $A_u$ |
| 1 | | | | |
| 2 | | | | |
| 3 | | | | |
| The average value of each tube $\bar{A}_{340}$ | | | | |

(2) Calculation (end-point method)

$$\text{serum glucose (mmol/L)} = \frac{A_U - A_C - A_B}{A_S - A_B} \times 5$$

**5**: concentration of standard glucose working solution(5 mmol/L).

(3) Normal reference value

Fasting serum glucose: 3.89-6.11 mmol/L(70-110 mg/dL).

### 2.4.2 Discussion

Why is hexokinase method more specific than glucose oxidase method?

*Zhou Ti*

# Chapter 15

## Extraction, Identification and Quantitative Analysis of Liver Glycogen

Glycogens mainly locate in skeletal muscle cells (2/3) and liver cells(1/3). The content of liver glycogen is nearly 5% of the total liver tissue, therefore, the liver is an ideal organ for this experiment.

## Experiment 1　Extraction, Identification and Quantitative Analysis of Liver Glycogen

## 1.1　Principle

### 1.1.1　Extraction and identification of liver glycogen

The glycogen can be released out from the liver cells after homogenized under trichloroacetic acid which can make protein to denature and precipitate. The glycogen is insoluble in ethanol and dissolved in hot water. 95% ethanol is used to precipitate the glycogen, then make the glycogen dissolved in hot water. The glycogen solution would show red and brown color in the presence of iodine. The glycogen has no reducibility, but can be hydrolyzed into reductive glucose by heating in an acidic solution. Glucose can make the divalent copper ion (Banshi's reagent) reduced being to monovalent copper precipitation. So the glycogen solution can be identified rapidly.

### 1.1.2　Determination of liver glycogen content

Glycogen is very stable in concentrated alkali solution. So glycogen extraction solution could be prepared after a small amount of liver tissue is first heated in concentrated alkali solution to destroy most of proteins and enzymes. Glycogen can be hydrolyzed into glucose in concentrated sulfuric acid (anthrone solution) which can further dehydrate the glucose into 5-(hydroxymethyl) furfural (HMF). Finally, a blue compound would be formed by subsequent dehydration and condensation reactions between HMF and anthrone. In the range of 10-100 μg, the blue color would be proportional to glucose concentration. Consequently, glycogen content can be determined by colorimetric analysis.

# 1.2 Materials

## 1.2.1 Reagents and solutions

(1) Sample: chicken liver.

(2) 95% ethanol.

(3) 5% trichloroacetic acid solution: dissolve 5 g trichloroacetic acid in distilled water to 100 mL.

(4) 12 mol/L HCl: concentrated HCl (36% –38%).

(5) 12.5 mol/L (50%) NaOH: dissolve 50 g NaOH in distilled water to 100 mL.

(6) Iodine reagents: dissolve 100 mg iodine and 200 mg KI in 50 mL distilled water.

(7) Banshi's reagent: dissolve 17.3 g sodium citrate ($C_5H_5Na_3O_7$ $5H_2O$, 294.10) and 100 g anhydrous sodium carbonate ($Na_2CO_3$ 105.99) in 700 mL distilled water, heat it to promote dissolving and then cool down, slowly pour into 100 mL of 17.3% copper sulfate ($CuSO_4 – 5H_2O$, 249.68) while shaking, qs to 1,000 mL with distilled water, mix well, filtrate if turbid. This reagent can be stored for a long time.

(8) 5.35 mol/L (30%) KOH solution: dissolve 30 g KOH in distilled water to 100 mL.

(9) Standard glucose solution (50 mg/L): dissolve 25 mg anhydrous glucose in distilled water to 500 mL.

(10) 17 mol/L (90%) $H_2SO_4$: slowly add 500 mL concentrated $H_2SO_4$ into 30 mL distilled water.

(11) 0.2% Anthrone reagent: dissolve 0.20 g anthrone in 17 mol/L $H_2SO_4$ to 100 mL. This reagent is unstable and should be freshing prepared. It can be stored in refrigerator at 4 ℃ for 4–5 days.

## 1.2.2 Special equipment

①Low – speed centrifuge; ②water bath; ③spectrophotometer; ④electronic balance (10 mg level); ⑤scissors; ⑥tweezers; ⑦mortar; ⑧white porcelain plate.

# 1.3 Methods

## 1.3.1 Extraction of liver glycogen

(1) Kill the chicken and take out the chicken liver immediately. Absorb the surface blood of liver tissue with filter paper.

(2) Weigh 1 g liver tissue with an electronic balance. Homogenize it into chyle with 1 mL of 5% trichloroacetic acid in the mortar.

(3) Add another 3 mL of 5% trichloroacetic acid, get the homogenate. Transfer all the homogenate to a clean 10 mL centrifuge tube. Then centrifuge at 4,000 r/min for 3 min.

(4) Transfer the supernatant (containing liver glycogen) to another 10 mL centrifuge tube.

(5) Add 5 mL of 95% ethanol and mix well, keep at room temperature for 10 min. The glycogen will be precipitated.

(6) Then centrifuge at 4,000 r/min for 5 min. Glycogen was thoroughly precipitated. Discard the supernatant and put the centrifuge tube inverted on the filter paper to dry for 2 min.

(7) Add 2 mL distilled water into a centrifuge tube and incubate it in boiling water for 2 min. Glycogen is dissolved and the glycogen solution is obtained.

## 1.3.2   Identification of liver glycogen

(1) Complete the following operations on the white porcelain plate according to the Table 15.1.

Table 15.1   Color reaction of glycogen

| Reagent(drop) | Sample | Blank |
| --- | --- | --- |
| Iodine reagents | 2 | 2 |
| Distilled water | – | 10 |
| Glycogen solution | 10 | – |

Mix well, observe the color reaction and record the results above.

(2) Prepare 2 test tubes and complete the following operations according to Table 15.2.

Table 15.2   Identification of glycogen

| Reagent(mL) | Sample | Blank |
| --- | --- | --- |
| Glycogen solution | 0.5 | 0.5 |
| HCl | 0.1 | – |
| Distilled water | – | 0.1 |
| | Boiling water bath for 10 min | Boiling water bath for 10 min |
| 50% NaOH | 0.1 | – |
| ddH$_2$O | – | 0.1 |
| Banshi's reagent(blue) | 1.5 | 1.5 |
| Colour | — | — |

Place in boiling water bath for 5 min. Observe the changes in color and precipitation, record the results.

## 1.3.3   Determination of liver glycogen content

(1) Add 1.5 mL of 30% KOH into a test tube, then weigh 0.10–0.15 g liver tissue (record the actual weight) with the electronic balance and put it into this tube. After incubating in boiling water bath for 15 min, and cooling down to roon temperature, transfer them to a 100 mL flask repeatedly. Wash the tube with distilled water and collect all of them into this flask, make sure that the final volume is exact 100 mL. This is the glycogen extraction solution.

(2) Prepare 3 test tubes and complete the following operations according to Table 15.3.

Table 15.3   Determination of liver glycogen

| Reagent(mL) | Blank | Standard | Sample |
| --- | --- | --- | --- |
| Distilled water | 1.0 | — | — |
| Standard glucose solution | — | 1.0 | — |
| Glycogen extraction solution | — | — | 1.0 |
| 0.2% Anthrone | 2.0 | 2.0 | 2.0 |

Mix well, place in boiling water bath for 10 min, read the absorbance at 620 nm by zeroing with the blank tube. Record the absorbance.

## 1.3.4 Notes

(1) The animals must be fed well before the experiment to avoid negative results due to glycogen degradation.

(2) When the liver is in vitro, it must be treated with trichloroacetic acid quickly to avoid the degradation of the glycogen by the enzyme.

(3) The grind should be sufficient to ensure the entire releasing of glycogen, otherwise, the whole experiment would be influenced.

(4) The sample tube must be adjusted to neutral or alkaline, otherwise, the function of the Banshi's reagent could be influenced.

(5) The liver tissue must be well suspended in the boiling water, otherwise, the absorbance value would be inaccurate.

(6) When calculating glycogen content according to the formula, take the actual liver tissue weight value into the formula.

(7) The anthrone reagent should be two times the volume of the tested solution. After adding anthrone reagent, it must be fully mixed.

# 1.4 Results and discussion

## 1.4.1 Results

(1) For the color reactions, sample tube would show brown–red color and blank tube ought to be original yellow of iodine. The brown–red color is due to the reactions between glycogen and iodine, however, these reactions do not occur in the blank tube.

(2) For the identification of glycogen, a few brick red precipitation could be observed at the bottom of the sample tube, but nothing could be observed in the blue ( color of Ban's reagent) blank tube. Concentrated hydrochloric acid would make the glycogen hydrolyzed into glucose. The reducibility of glucose could make the divalent copper ion ( Banshi's reagent) reduced to monovalent oxidized cuprous precipitation. However, there is no hydrochloric acid in blank tube and glucose could not be produced. Consequently, no brick red precipitation could be observed.

(3) Calculate the liver glycogen content according to the following equation 15. 1.

$$\text{Liver glycogen content ( g/100 g Liver)} = \frac{A_{sample}}{A_{standard}} \times 0.05 \times \frac{100}{\text{tissue weight( g)}} \times \frac{100}{1,000} \times 1.11$$

( Equation 15. 1)

*Note: in the color reaction of anthrone, the effect of* 100 *μg glucose is equivalent to that of* 110 *μg glycogen. So* 1. 11 *is the constant that converting glucose content to that of glycogen.*

This method is only suitable for liver glycogen content between 1. 5% –9. 0% . If the glycogen content of the liver is <1. 0% , the anthrone reaction could be interfered by protein, therefore, an indirect method would be better. After digesting the liver tissue with concentrated alkali solution, the glycogen could be precipitated by 95% ethanol first ( concentrated alkali: 95% ethanol =1 ∶ 1. 25), and then be heated with 2 mL distilled water to dissolve them all. Finally, anthrone reagent could be applied for colorimetric analysis and calculation.

## 1.4.2　Discussion

(1) What is the role of trichloroacetic acid in the extraction of glycogen?

(2) For extraction of liver glycogen, what reagents could precipitate the glycogen?

(3) Why it shows brown red color when adding iodine to glycogen? How the difference compared to starch?

(4) What could be produced when glycogen is hydrolyzed by concentrated hydrochloric acid? What is the experiment phenomenon after Banshi's reagent test?

(5) What are the main components of anthrone reagent and their effects?

(6) Without anthrone reagent, can you redesign a new experimental scheme to measure glycogen content?

*Wu Yanhui*

# Chapter 16

## Determination of Serum Lipids

Lipids are a chemically diverse group of organic molecules, the defining property of which is water insolubility. Lipids include fats and lipoids, their biological functions are diverse. Fats (i. e. , triglycerides) are the major stored energy source in many organisms. Lipoids refer to other lipids except for fats, including phospholipids, sterols, glycolipids, etc. Although lipids present in relatively small quantities, they play a series of important biological functions. The phospholipids and steroids are the main structural elements of biological membranes. Cholesterol can be converted to bile acids in liver and excrete into bile to participate in lipids digestion. The hydrolysis products of phosphatidyl inositol (i. e. , DAG, $IP_3$) are important secondary messengers in cell signal transduction.

Plasma lipids refer to all the lipids exist in the blood plasma, including triglycerides, phospholipids, cholesterol and cholesteryl esters. As water-insoluble compounds, lipids are transfered in the blood plasma as plasma lipoproteins by specific carrier proteins called apolipoproteins. Hyperlipidemia is usually caused by lipids metabolic disorders, and dysregulation of cholesterol metabolism can lead to cardiovascular disease. In this chapter, the methods of triglycerides and cholesterol determination will be discussed.

## Experiment 1　Determination of Triglyceride in Serum by Acetylacetone Method

## 1. 1　Principle

Serum triglyceride is isolated by n-heptane/isopropanol mix and saponified by KOH, and then oxidized by periodic acid to formaldehyde; methanal reacts with acetylacetone agent to produce a yellow derivative of dihydro-dimethylpyridine. Run colorimetric assay at 420 nm and calculate with the standard solution treated in the same way to obtain the concentration of serum triglyceride.

(1) Saponification

$$R-\overset{\overset{O}{\|}}{C}-O-\overset{\overset{\displaystyle CH_2-O-\overset{\overset{O}{\|}}{C}-R}{|}}{\underset{\underset{CH_2-O-\overset{}{\underset{\|}{O}}-R}{|}}{CH}} + 3KOH \longrightarrow 3R-\overset{\overset{O}{\|}}{C}-OK + \overset{\displaystyle CH_2OH}{\underset{\underset{CH_2OH}{|}}{\overset{|}{CHOH}}}$$

(2) Oxidation of glycerol

$$\begin{array}{c} CH_2OH \\ | \\ CHOH \\ | \\ CH_2OH \end{array} + 2HIO_4 \longrightarrow 2HCO + HCOOH + 2HIO_3 + H_2O$$

(3) The reaction of formaldehyde with acetylacetone

$$HCHO + 2CH_3-\overset{O}{\overset{\|}{C}}-CH_2-\overset{O}{\overset{\|}{C}}-CH_3 + NH_4^+ \xrightarrow{-3H_2O}$$

# 1.2    Materials

## 1.2.1    Reagents and solutions

(1) Isolation solution: n-heptane/isopropanol = 2 : 3.5(v/v)

(2) 0.04 mol/L $H_2SO_4$.

(3) Isopropanol.

(4) Saponifier: dissolve 6 g KOH in 60 mL distilled water, add 40 mL isopropanol, mix well, store in a brown bottle at room temperature.

(5) Oxidant: dissolve 650 mg sodium periodate in 500 mL distilled water, add 77 g anhydrous ammonium acetate, after completely dissolved, add 60 mL glacial acetic acid and dilute to 1,000 mL with distilled water, store in a brown bottle at room temperature.

(6) Acetylacetone agent: take 0.4 mL acetylacetone, add isopropanol to 100 mL; place in a brown bottle.

(7) Standard stock solution (10 g/L): dissolve 1.0 g triglycerides in isolation solution to 100 mL; mix well, store in a brown bottle.

(8) Standard working solution (1.0 g/L): take 10 mL standard stock solution, add 90 mL isolation solution, store in the refrigerator.

## 1.2.2    Special equipment

①Spectrophotometer; ②thermostat water bath; ③pippet; ④glass tubes.

# 1.3    Methods

## 1.3.1    Isolation

Label 3 dry small tubes, add reagents according to Table 16.1.

Table 16.1    Isolation system

| Reagents ( mL) | Blank tube | Standard tube | Sample tube |
|---|---|---|---|
| Serum | — | — | 0.2 |
| Standardworking solution( 1.0 g/L) | — | 0.2 | — |
| Distilled water | 0.2 | — | — |
| Isolation solution | 2.0 | 2.0 | 2.0 |
| 0.04 mol/L $H_2SO_4$ | 0.6 | 0.6 | 0.6 |

Shake the tube while adding, shake vigorous for 5 min after adding; place them to layer.

### 1.3.2    Saponification

(1) Remove 0.5 mL supernatant respectively to 3 large tubes with the same label ( ensure no lower layer solution is removed), add reagents according to Table 16.2.

Table 16.2    Saponification system

| Reagents ( mL) | Blank tube | Standard tube | Sample tube |
|---|---|---|---|
| Supernatant | 0.5 | 0.5 | 0.5 |
| Isopropanol | 2.0 | 2.0 | 2.0 |
| Saponifier | 0.4 | 0.4 | 0.4 |

(2) Mix immediately, incubate in water bath at 65 ℃ for 5 min.

(3) Oxidation and color reaction: add 2.0 mL oxidant to the tubes above, mix well and add 2.0 mL acetylacetone; shake completely and place in water bath at 65 ℃ for 15 min; then cool down to room temperature, zero the instrument with blank tube and measure the absorbance of each tube at 420 nm.

*Notes:* ①*To ensure the coincidence of the reaction condition in the process of operation and oxidation, prepare standard tube each time to reduce error. If the* $A_{420}$ *value is higher than 0.7, the amount of serum should be half.* ②*The sample with hemolysis has a small influence.*

## 1.4    Results and discussion

### 1.4.1    Data record

Record the absorbance of each tube at 420 nm ( Table 16.3).

Table 16.3    Absorbance data

| Tests | Absorbance( $A_{420}$ ) | |
|---|---|---|
| | Standard tube | Sample tube |
| 1 | | |
| 2 | | |
| 3 | | |
| Average $\overline{A}_{420}$ | | |

## 1.4.2　Calculation

$$\text{Serum triglyceride (mmol/L)} = \frac{\text{Absorbance of sample tube}}{\text{Absorbance of standard tube}} \times 1.0 \times \frac{1,000}{886}$$

1.0: concentration of standard working solution (1.0 g/L). **886**: average molecular weight of triglyceride. Normal values: the normal value is 0.34−1.70 mmol/L in young adults.

## 1.4.3　Discussion

①What are the effects of isolation solution (n−heptane/isopropanol) in this experiment? ②What is the purpose of saponification? What problems should be paid attention to obtain the precise amount of serum triglyceride?

# Experiment 2　Determination of Cholesterol in Serum by Cholesterol Oxidase Method

There are two forms of cholesterol in the body include both the free form cholesterol and the form cholesterol ester. Cholesterol is an important component of the lipoproteins, and it is also required in the membrane of the mammalian cell for maintenance of cellular functions. Moreover, cholesterol is the common precursor of other steroids. It can be converted to bile acids, steroid hormones, and vitamin $D_3$. Only less than 20% of the cholesterol in the blood is acquired from the diet, the rest of the cholesterol is primarily synthesized in the cytosol and endoplasmic reticulum of many cells. The cholesterol is continuous to be metabolized in the blood circulation system through exchanging with other tissues. Therefore, detecting the cholesterol in the serum can help us know the synthesis, transportation, and uptake of cholesterol in the body. The serum cholesterol value can guide the prevention of diseases such as atherosclerosis. The normal value of serum cholesterol in adults is 100−250 mg/dL (2.59−6.47 mmol/L).

Serum cholesterol detection can be performed by either chemical method or an enzymatic method. The chemical method including 5 major steps: extraction, saponification, purification, chromogenic reaction and colorimetric assay. The enzymatic method can detect total serum cholesterol through the combined utilization of cholesterol esterase, cholesterol oxidase, and peroxidase. The enzymatic determination is a common assay of serum cholesterol recommended by the Chinese Medical Association. The enzymatic assay is a fast and accurate method exhibited several advantages including improved specificity, smaller sample volume, and rapid sample throughput. Moreover, the enzymatic method can be used to detect free cholesterol directly if the cholesterol esterase component is omitted.

## 2.1　Principle

The cholesteryl ester (CE) is hydrolyzed to free cholesterol (FC) by cholesterol esterase (CHER). Then FC is oxidized by cholesterol oxidase (CHOD) to form the corresponding cholest−4−en−3−one and hydrogen peroxide ($H_2O_2$). Thus, the enzymatic method for determining FC is based on the measurement of hydrogen peroxide. The hydrogen peroxide reacts with 4−aminoantipyrine (4−AAP) and phenol to form a red product, quinine derivative (Trinder reaction). The concentration of cholesterol can be determined by measuring the absorbance of the product at 500 nm.

Only the free cholesterol in the serum can be oxidized to form hydrogen peroxide for further detection if cholesterol esterase is excluded. Therefore, the enzymatic method is also a simple way for the direct detection of free cholesterol in the serum.

$$\text{Cholesteryl ester} + H_2O \xrightarrow{\text{Cholesterol esterase}} \text{Cholesterol} + \text{free fatty acids}$$

$$\text{Cholesterol} + O_2 \xrightarrow{\text{Cholesteroloxidase}} \text{Cholestenone} + H_2O_2$$

$$H_2O_2 + \text{Phenol} + 4-\text{amino}-\text{antipyrene} \xrightarrow{\text{Peroxidase}} \text{Quinone derivertives} + H_2O$$

## 2.2 Materials

### 2.2.1 Reagents and solutions

(1) Cholesterol enzymatic assay kit (Table 16.4): 20 mL enzyme mix, 80 mL dilution buffer.

Table 16.4 **Components in the kit**

| Reagent | Concentration |
|---|---|
| GOOD's buffer (pH 6.7) | 50 mmol/L |
| Cholesterol esterase | $\geqslant$ 200 U/L |
| Cholesterol oxidase | $\geqslant$ 100 U/L |
| Hydrogen peroxidase | $\geqslant$ 3,000 U/L |
| 4-AAP | 0.3 mmol/L |
| Phenol | 5 mmol/L |

(2) The cholesterol esterase can be omitted if free cholesterolis needed to be determined in the sample.

(3) Dilution buffer: 50 mmol/L GOOD's buffer (pH 6.7)

(4) Enzyme working solution: dilute the enzyme mix with dilution buffer (1 : 4).

(5) Standard cholesterol solution (5.17 mmol/L): weigh 200 mg cholesterol exactly; place in a volumetric flask; add isopropanol to 100 mL (2 mg/mL), store the standard solution in the refrigerator at 4 ℃.

### 2.2.2 Special equipment

①Spectrophotometer; ②thermostat water bath; ③pippet; ④Glass tubes.

## 2.3 Methods

### 2.3.1 Reaction system

Label 3 dry small tubes, add reagents according to Table 16.5.

Table 16.5 **Reaction system setting**

| Reagent(mL) | Blank tube | Standard tube | Sample tube |
|---|---|---|---|
| Serum | — | — | 0.04 |
| Standard cholesterol solution | — | 0.04 | — |
| Distilled water | 0.04 | — | — |
| Enzyme working solution | 3.0 | 3.0 | 3.0 |

## 2.3.2   Reaction

Mix the solution and then incubate in water bath at 37 ℃ for 15 min.

## 2.3.3   Detection

Zero the instrument with blank tube and measure the absorbance of each tube at 500 nm. Repeat detection of each tube for three times.

## 2.3.4   Notes

(1) The quality of the enzyme has a significant effect on the final result. The enzyme mix with high quality can catalyze the reaction within 5 min, however, enzyme mix with poor quality may lower the final result.

(2) In the presence of cholesterol esterase, the assay determines total cholesterol. To detect free cholesterol only, omit the cholesterol esterase from the enzyme mix. Normally, free cholesterol accounts for about one-third of total cholesterol.

(3) The enzyme mix can be stored in the dark for 6 months at 2-8 ℃. However, the enzyme working solution can only store for 7 days after preparation. Make sure the reaction time of each tube is exactly the same during the experiment.

(4) Normally, the cholesterol is determined in the serum. Plasma sample can also be used for the determination, however, anticoagulant EDTA may reduce the total cholesterol concentration dose-dependently.

(5) The total cholesterol in the serum remains within normal limits even there is mild hemolysis or the sample contains a small amount of bilirubin. However, severe hemolysis is supposed to cause increased results.

(6) Detection of serum cholesterol has important clinical significance. Increases of total serum cholesterol levels are found most characteristically in the primary hyperlipidemia, atherosclerosis, diabetes mellitus, nephrotic syndrome, bile ducts obstruction and hypopituitarism. Reduced level of cholesterol occurs in hyperthyroidism, hepatic disease, severe infection, malnutrition and hemolytic jaundice caused by drug treatment.

# 2.4   Results and discussion

## 2.4.1   Data record

Record the absorbance of each tube at 500 nm (Table 16.6).

Table 16.6   **Absorbance data**

| Tests | Absorbance($A_{500}$) | |
| --- | --- | --- |
| | Standard tube | Sample tube |
| 1 | | |
| 2 | | |
| 3 | | |
| Average $\bar{A}_{500}$ | | |

## 2.4.2 Calculation

$$\text{Serum cholesterol ( mmol/L)} = \frac{\text{Absorbance of sample tube}}{\text{Absorbance of standard tube}} \times 5.17$$

**5.17**: concentration of standard cholesterol solution (5.17 mmol/L).

## 2.4.3 Discussion

(1) What kind of factors may affect the final result of the determination?

(2) How would the result be affected if the sample was collected from a serious jaundice patient?

(3) List the widely used cholesterol detection methods and compare the advantages and disadvantages of each method.

# Experiment 3   Determination of High-density Lipoprotein-cholesterol in Serum

## 3.1 Principles

Lipoprotein in serum can bind to divalent metal ions and form insoluble complexes. The forming of these complexes is correlated to the ratio of protein/lipid( P/L) in lipoprotein. The smaller P/L ratio, the easier to precipitate. Therefore the right concentration of magnesium −phosphotungstate can precipitate CM, LDL, VLDL, and leave only HDL in the supernatant.

Use o−phthalaldehyde−glacial acetic acid−$H_2SO_4$ for color reaction of cholesterol, compared with standard cholesterol solution, the content of serum HDL−cholesterol can be determined by colorimetry.

## 3.2 Materials

### 3.2.1 Reagents and solutions

(1) Samples: serum samples from fasting healthy persons.

(2) Magnesium−phosphotungstate ( Mg−OPA).

(3) Dissolve 4 g phosphotungstic acid in 50 mL ddH$_2$O, add 16 mL of 1 mol/L NaOH, mix, which is add ddH$_2$O to 100 mL, adjust pH to 7.6. Take 40 mL the mixture, which is add 10 mL of 2 mol/L MgCl$_2$, then add normal saline to 100 mL.

(4) o−Phthalaldehyde solution.

(5) Stock solution (100 mg/dL): dissolve 100 mg o−phthalaldehyde in 100 mL glacial acetic acid, store in a brown bottle.

(6) Working solution (5 mg/dL): take 50 mL storage solution, add 20 mL absolute ethanol and 20 mL ethyl acetate, mix and add glacial aceticacid to 1,000 mL, store in a brown bottle.

(7) Concentrated sulfuric acid ( AR).

(8) Normal saline: 0.9% ( w/v) NaCl.

(9) Standard cholesterol solution (40 mg/dL): dissolve 40 mg cholesterol in 100 mL glacial acetic acid.

## 3.2.2   Special equipment

①722-Spectrophotometer; ②centrifuge and centrifuge tubes.

# 3.3   Methods

## 3.3.1   Separation of HDL

Take a test tube, add 0.2 mL serum, mix with 0.05 mL Mg-OPA, place at room temperature for 10 min. Centrifuge at 2,500 r/min for 10 min. CM, LDL, VLDL will precipitate after centrifugation and only HDL remain in the supernatant.

## 3.3.2   Measurement of HDL-cholesterol

Label three tubes, add reagents according to the Table 16.7.

Table 16.7   Reaction system setting

| Reagent(mL) | Blank tube | Standard tube | Sample tube |
|---|---|---|---|
| Supernatant with HDL | – | – | 0.05 |
| Standard cholesterol solution | – | 0.05 | – |
| Normal saline | 0.05 | – | – |
| o-Phthalaldehyde | 2.5 | 2.5 | 2.5 |
| Concentrated H$_2$SO$_4$ | 1.5 | 1.5 | 1.5 |

Mix each tube, zero the instrument with blank tube and measure the absorbance of each tube at 550 nm, record $A_{550}$ for standard and sample tubes.

## 3.3.3   Calculation

HDL-cholesterol Conc. $= A_{sample}/A_{st\,standard} \times$ Standard chol Conc. ( mmol/L or mg/dl) ×2 *

* is the serum dilution factor.

## 3.3.4   Notes

(1) The supernatant should be clear and transparent, contain no granule. Otherwise, re-centrifuge is required.

(2) The test tube must tilt when adding concentrated sulfuric acid, add liquid along the tube wall and take great care.

(3) The stable time of color reaction is about 1 h. Detect the absorbance in 1 h.

# 3.4   Results and discussions

## 3.4.1   Results

The normal range of HDL-C determined by this method is 45-90 mg/dL or 1.17-2.33 mmol/L.

## 3.4.2   Discussions

(1) How to reduce the measurement errors in the HDL-C determination?

(2) What is the clinic significance for high HDL-C or low HDL-C?

# Experiment 4   Separation of Serum Lipoproteins by Agarose Gel Electrophoresis

## 4.1   Principles

The lipids in serum combine with hydrophilic apolipoproteins to form lipoproteins. Different lipoproteins contain different kinds and amounts of apolipoproteins. Their sizes and dense also have great difference. Agarose gel electrophoresis can easily separate the serum lipoprotein particles. This method is simple and inexpensive, therefore it is widely used in the lab.

Using lipid dyes such as Sudan black B or oil red O to prestain the serum lipoproteins, then separate the stained serum by agarose gel electrophoresis. Since the net charges of all lipoproteins in the pH 8.6 buffer system are negative, the lipoproteins will all move toward the anode. Then lipoproteins will be separated into different bands according to their mass to charge ratios.

## 4.2   Materials

### 4.2.1   Reagents and solutions

(1) Samples: serum samples from fasting healthy persons.

(2) Barbital buffer solution (BBS, pH 8.6, ionic strength 0.075): dissolve 15.5 g Barbital-Na, 2.76 g barbital, 0.292 g EDTA in distilled water to 1,000 mL.

(3) Tris buffer (pH 8.6): dissolve 1.212 g Tris, 0.29 g EDTA, 15.85 g NaCl in distilled water to 1,000 mL.

(4) Agarose gel: add 0.5 g agarose, 50 mL Tris buffer, 50 mL distilled water, heat to the boiling point, stop heating when agarose is melted.

(5) Sudan black staining solution: add Sudan black B to absolute ethanol until saturation, filter before using.

### 4.2.2   Special equipment

①Knife blade electrophoresis apparatus; ②electrophoresis chamber.

## 4.3   Methods

### 4.3.1   Protocol A

(1) Prestain of serum: mix 0.2 mL serum with 0.2 mL Sudan black B staining solution in a small test

tube, incubate in 37 ℃ water bath for 30 min, then centrifuge at 2,000 r/min for 5 min, get rid of the staining debris in the serum.

(2) Casting agarose gel plate: take a clean glass slide, melt 0. 5% agarose gel by the microwave oven and add 2. 5 mL to the slide, wait 30 min for the gel to be solidified (the gel slide can be put into the fridge to speed up it's cooling in summer).

(3) Loading sample: when the gel is a complete set, mark the slide at 2 cm from one end, use the blade to cut a small thin well, add 15 mL pre-stained serum in the well.

(4) Electrophoresis: put the gel slide with the serum sample in the electrophoresis chamber, the sample is at the cathode side. Dip two pieces of three-layer gauze into BBS and stick them to the two ends of the gel slide (*Note: the BBS in the electrophoresis chamber can not be replaced by Tris buffer*). Switch on the power, the constant voltage at 120 – 130 V, 3 – 4 mA for one gel slide, run for 15 – 55 min, the separated bands can be visible.

## 4.3.2　Protocol B

(1) Prestain of serum: mix 0. 2 mL serum with 0. 2 mL Sudan black B staining solution in a small test tube, incubate in 37 ℃ water bath for 30 min, then centrifuge at 2,000 r/min for 5 min, get rid of the staining debris in the serum.

(2) Casting agarose gel plate: take a clean glass slide, melt 0. 5% agarose gel by microwave oven and add 2. 5 mL to the slide, wait 30 min for the gel be solidified (the gel slide can be put into the fridge to speed up it's cooling in summer).

(3) Loading sample: when the gel is a complete set, mark the slide at 2 cm from one end. Cut a filter paper and fold it, use the folded filter paper to cut a sample application point at 2 cm from one end, use a capillary to sip the prestained serum and apply it to the point for three seconds.

(4) Electrophoresis: put the gel slide with the serum sample in the electrophoresis chamber, the sample is at the cathode side. Dip two pieces of gauze into BBS to form bridges and stick them 1 cm to two ends of the gel slide. Switch on the power, 3 – 4 mA per gel slide run for 10 – 15 min, then switch to 6 – 7 mA per slide and run for 30 – 40 min, the separated bands can be visible.

(5) Storage: to preserve the electropherogram, put the gel slide in water for 2 hours to desalt, and then put in oven bake at 80 ℃.

## 4.3.3　Notes

(1) The sample for electrophoresis should be fresh, fasting serum.

(2) Each gel slide can be cutted to two wells, therefore can load two samples.

(3) Instead of cutting wells, the comb can also be used when casting the gel.

(4) Avoid losing too much water during melt the gel. It is better to use freshly prepared gel.

(5) Use 0. 5% agarose gels are usually good for separation of serum lipoprotein. Higher than 1% will cause bad separation between β and pre-β-lipoproteins. Lower than 0. 45% will cause the gel to become fragile and also reduce the resolution.

(6) Sample loading wells should be cut neatly and smoothly, otherwise may affect the electrophoresis bands.

(7) If there is a shallow band appeared in front of α-lipoprotein, it can be considered as pre α-lipoprotein.

# 4.4   Results and discussions

## 4.4.1   Results

From cathode to anode, the fast serum samples from healthy people will show three lipoprotein bands: β−lipoprotein ( darkest) , pre−β−lipoprotein and α−lipoprotein ( darker than pre−β−lipoprotein) .

## 4.4.2   Discussions

(1) Why the serum samples from fasting healthy people only show three lipoprotein bands from the cathode to anode? i. e. , β−lipoprotein ( darkest) , pre−β−lipoprotein and α−lipoprotein ( darker than pre−β −lipoprotein) ?

(2) If the pre−β−lipoprotein is darker than α−lipoprotein, and TG ( TAG) in the serum increased significantly while TC ( total cholesterol) in the serum remains normal or slightly higher, what kind of hyperlipoproteinemia can we conclude?

(3) If the β−lipoprotein is darker than the normal serum, and TC in the serum increased significantly while TG in the serum remains normal, what kind of hyperlipoproteinemia can we conclude?

(4) If the β−lipoprotein is darker than the normal serum, TC in the serum increased and TG in the serum also increase, what kind of hyperlipoproteinemia can we conclude?

(5) If β−lipoprotein cannot be separated from pre−β−lipoprotein, fused as a broader β−band, and TC, TG in the serum all increased, what kind of hyperlipoproteinemia can we conclude?

(6) If CM appear in the loading point, β− and pre−β−lipoprotein remain normal or slightly decreased, while TC increased significantly, what kind of hyperlipoproteinemia can we conclude?

*Wang Xuejun, Yin Ye, Wang Yanfei*

# Chapter 17

## Determination of Blood Glucose and Serum Lipids

Determination of blood glucose and lipid is the major practice of clinical biochemistry. In this chapter, the experiment is linked with disease and research by constructing a diabetic rat model. By observing the effect of the hormone on blood glucose, triglyceride and cholesterol of normal rat and diabetic rat, we will understand the difference of carbohydrate and lipid metabolism in normal rat and diabetic rat, and the mechanism and clinical significance of blood glucose and lipid assay.

## Experiment 1 The Effect of the Hormone on Blood Glucose and Lipids of Normal and Diabetic Rat

### 1.1 Principle

Streptozocin (STZ) is a cellular toxic compound. It can be selectively targeted to the β cell of the pancreas and induce type 1 diabetes by causing DNA damage. So STZ can be used to construct a diabetic rat model, then the blood glucose and lipid levels in control rats and diabetic rats induced by STZ can be compared in this experiment.

Glucose oxidase (GOD) method is used extensively to determine blood glucose inthe clinic because of its high specificity, low cost, and convenient operation (See Experiment 1, Chapter 14).

Triglyceride is oxidized by lipoprotein lipase to produce glycerol and fatty acids. Glycerol is further phosphorylated into 3−phosphoglycerol by glycerol kinase. 3−phosphoglycerol is then oxidized by glycerol-phosphate oxidase to produce $H_2O_2$. $H_2O_2$ is then oxidatively coupled with 4−APP and EHSPT [ N−ethyl−N−(2−hydroxy−3−sulfopropyl) 3−m−toluidine] in the presence of POD to yield quinone derivatives that is measured at 546 nm. The absorbance at 546 nm is proportional to the triglyceride concentration in the sample.

$$\text{Triglyceride} + 3H_2O \xrightarrow{\text{Lipoprotein lipase}} \text{Glycerol} + 3 \text{ Fatty acids}$$

$$\text{Glycerol} + \text{ATP} \xrightarrow{\text{Glycerol kinase}} \text{Glycerol}-3-\text{phosphate} + \text{ADP}$$

$$\text{Glycerol}-3-\text{phosphate} + O_2 \xrightarrow{\text{Phosphoglycerol oxidase}} \text{Dihydroxyacetone phosphate} + H_2O_2$$

$$\text{EHSPT} + H_2O_2 + 4-\text{amino}-\text{antipyrene} \xrightarrow{\text{Peroxidase}} \text{Quinone derivertives} + H_2O$$

Determination of cholesterol in serum can be performed by enzymatic method (See Experiment 2, Chapter 16).

# 1.2 Materials

## 1.2.1 Animals

Adult male SD(Sprague-Dawley) rats weighing about 200 g.

## 1.2.2 Reagents and solutions

(1)Standard glucose solution (5.05 mmol/L), standard cholesterol solution (5.17 mmol/L) and standard triglyceride solution (2.26 mmol/L): commercial reagents are recommended.

(2)Medical insulin (40 IU/mL).

(3)Medical adrenaline (1 mg/mL).

(4)Blood glucose assay reagent (kit): glucose oxidase >10 U/mL, peroxidase > 1 U/mL, phosphate 70 mmol/L, phenol 5 mmol/L, 4-amino-antipyrene: 0.4 mmol/L, pH 7.0.

(5)Total cholesterol assay reagent (kit): pipes [piperazine − N, N′ − bis(2 − ethanesulfonic acid)] 35 mmol/L, cholesterol oxidase>0.1 U/mL, phenol 28 mmol/L, sodium cholate 0.5 mmol/L, 4−AAP 0.5 mmol/L, cholesterol esterase >0.2 U/mL, peroxidase >0.8 U/mL, pH 7.0. Commercial kit is recommended.

(6)Triglyceride assay reagent (kit): pipes 45 mmol/L, magnesium chloride 5.0 mmol/L, glycerol kinase >1.5 U/mL, lipoprotein lipase >100 U/mL, 3−phosphate glycerol oxidase >4 U/mL, EHSPT [N−Ethyl−N−(2−hydroxy−3−sulfopropyl) −3−methylaniline] 3.0 mmol/L, 4−APP 0.75 mmol/L, peroxidase >0.8 U/mL, ATP 0.9 mmol/L, pH 7.5. Commercial kit is recommended.

(7)0.05 mol/L citric acid solution: dissolve 1.050,7 g citric acid in 100 mL ddH$_2$O and adjust pH to 4.5. Make a filtrated sterilization by 0.22 μm filter paper.

(8)STZ (Sigma): dissolve an appropriate amount of STZ in 0.05 mol/L citric acid solution (such as 18 mg/mL). STZ should be freshly prepared on ice and can not be stored for a long time.

## 1.2.3 Special equipment

①Centrifuge; ②eppendorf tube; ③pipette.

# 1.3 Methods

## 1.3.1 Procedure

(1)Construction of the diabetic rat model

Rats are divided into a control group and the diabetic group. The diabetic group is treated with STZ by intraperitoneal injection at a dose of 60 mg/kg weight mass. The control group is treated with the same volume of citric acid solution. After 24, 48 and 72 h of injection, blood glucose is measured respectively. Those rats with fasting blood glucose greater than 13.8 mmol/L (see Troubleshooting for normal reference) are identified as diabetic rats (the step can be omitted due to the stability and maturation of the method). Rats fast 16 h befere of the experiment.

(2) Tail bleeding

The tail of rat is firstly warmed by immersing into water bath at 45 ℃ (the tail also can be spread with dimethylbenzene). Then the tail is cleaned with a cotton swab moistened with 70% ethanol. Snip off the very tip of the rat's tail (5 mm to the end). The rat's tail should be gently squeezed, starting at the base of the tail and moving towards the site of the cut, to get a drop of blood from the incision site. The drops of blood should be carefully brought into the tubes and mark the sample with "Pre" (means before hormone injection). 200–300 μL blood is enough. Then the cut should be staunched by compressing.

(3) Administering hormone injection

Mark the control rats and diabetic rats. Both groups of rats are further divided into two sub–groups marked with group 1 and group 2. Group 1 is injected with insulin, 0. 75 IU/kg (60 U/kg), taking 200–300 μL of blood into the tube after 30 min injection and mark with "Ins". Group 2 is injected with adrenaline, 0. 1% adrenaline (0. 2 mg/kg), 30 min later, taking 200–300 μL of blood into the tube, mark with "Adr".

Separate the serum by centrifugation at 3,000 r/min for 5 min and take the supernatant for the following determination.

(4) Blood glucose determination (GOD–POD method)

Label 5 tubes, add reagents according to Table 17. 1.

Table 17.1    Blood glucose assay

| Reagent ( μL) | Blank | Pre * | Ins * | Adr * | Standard |
|---|---|---|---|---|---|
| Serum | – | 10 | 10 | 10 | – |
| Standard glucose solution (5. 05 mmol/L) | – | – | – | – | 10 |
| dH$_2$O | 10 | | | | |
| Blood glucose assay reagent | 1,000 | 1,000 | 1,000 | 1,000 | 1,000 |

∗ Pre: the group before hormone injection. ∗ Hormone (insulin or adrenaline) injected group contains both control rats and diabetic rats.

Mix the tubes well and incubated at 37 ℃ for 10 min. Blank the spectrophotometer at 505 nm with blank tube (Blank). Record the $A_{505}$ of each tube.

(5) Triglyceride determination (colorimetric method)

Label 5 tubes, add reagents according to Table 17. 2.

Table 17.2    Blood triglyceride assay

| Reagent ( μL) | Blank | Pre | Ins | Adr | Standard |
|---|---|---|---|---|---|
| Serum | – | 10 | 10 | 10 | – |
| Standard triglyceride solution (2. 26 mmol/L) | – | – | – | – | 10 |
| ddH$_2$O | 10 | | | | |
| Triglyceride assay reagent R1 | 800 | 800 | 800 | 800 | 800 |
| | | Mix and incubate at 37 °C for 5 min | | | |
| Triglyceride assay reagent R2 | 200 | 200 | 200 | 200 | 200 |

Mix the tubes well and incubated at 37 ℃ for 5 min. Blank the spectrophotometer at 546 nm with blank tube (Blank). Record the $A_{546}$ of each tube.

(6) Total cholesterol determination (colorimetric method)

Label 5 tubes, add reagents according to Table 17.3.

Table 17.3   Blood cholesterol assay

| Reagent ( μL) | Blank | Pre | Ins | Adr | Standard |
|---|---|---|---|---|---|
| Serum | – | 10 | 10 | 10 | – |
| Standard cholesterol solution (5.17 mmol/L) | – | – | – | – | 10 |
| dH$_2$O | 10 | | | | |
| Cholesterol assay reagent | 1,000 | 1,000 | 1,000 | 1,000 | 1,000 |

Mix the tubes well and incubated at 37 ℃ for 10 min. Zero the spectrophotometer at 505 nm with blank tube (Blank). Record the $A_{505}$ of each tube.

## 1.4   Results and discussion

### 1.4.1   Calculation

(1) Using the data obtained from the tables above to calculate the blood glucose and lipid according to the following formula.

$$Blood\ sugar/lipid(mmol/L) = \frac{A_{sample}}{A_{standard}} \times Concentration\ of\ standard$$

References value of rat serum/plasma:

Blood glucose: 2.64–5.26 mmol/L.

Blood cholesterol: 1.0–1.5 mmol/L.

Blood triglyceride: 0.4–0.7 mmol/L.

(2) Collect the data from different groups and fill in the following Table 17.4 and take an overview of the change of indexes between control and diabetic mellitus rats.

Table 17.4   Summary sheet of data from different groups

| Treatment index | Construct of DM rat | | 1 week after the construction of DM * | | | | | |
|---|---|---|---|---|---|---|---|---|
| | Control | DM | Pre | | Ins | | Adr | |
| | | | Control | DM | Control | DM | Control | DM |
| Blood glucose (mmol/L) | | | | | | | | |
| | | | | | | | | |
| | | | | | | | | |
| | | | | | | | | |
| | | | | | | | | |
| | | | | | | | | |

Continue to Table 17.4

| Treatment Index | Construct of DM rat | | 1 week after the construction of DM * | | | | | |
|---|---|---|---|---|---|---|---|---|
| | Control | DM | Pre | | Ins | | Adr | |
| | | | Control | DM | Control | DM | Control | DM |
| Blood triglyceride (mmol/L) | | | | | | | | |
| | | | | | | | | |
| | | | | | | | | |
| | | | | | | | | |
| | | | | | | | | |
| | | | | | | | | |
| Blood cholesterol (mmol/L) | | | | | | | | |
| | | | | | | | | |
| | | | | | | | | |
| | | | | | | | | |
| | | | | | | | | |
| | | | | | | | | |

\* Establish DM (diabetic mellitus) rat at first, one week later, determine blood glucose and lipid of rats.

All the data are collected in Table 17.4 for further comparison and statistic.

## 1.4.2　Discussion

Explain the mechanism of insulin in regulating blood glucose and lipids linking with diabetes mellitus.

# 1.5　Troubleshooting

(1) If plasma is used for detection, the syringe and tubes should be properly moistened with heparin before being used or use anticoagulative tubes directly.

(2) STZ solution should be put on the ice and used after it is ready and cannot be stored for a long time.

(3) The rat is too big for beginners to operate and muffler should be worn to avoid bitten by rats.

(4) The diabetic rat is under weight loss and disease condition. It is sometimes difficult to do tail bleeding for diabetic rats. Doing much more exercise can increase blood circulation of diabetic rats and benefit tail bleeding. A glucometer is recommended for some microscale sample assay.

*Li Jiao, Lü Lixia*

# Nucleic Acid Experiments

# Chapter 18

## Isolation, Quantitative Analysis and Identification of DNA

To isolate, purify, quantify and identify enough quality plasmid DNA and genomic DNA from all kinds of species samples are the basic techniques of molecular biology experiments.

## Experiment 1   Isolation, Quantification and Identification of Plasmid DNA

Plasmid DNA purification is the routine technique applied in gene cloning. Many methods have been developed to purify plasmid DNA from bacteria, including alkaline lysis method, boiling method, hydroxyapatite chromatography, CsCl−ethidium bromide density gradient centrifugation, column centrifugation and so on. Among them, alkaline lysis method, the most classical and common one, is suitable for the isolation of plasmid DNA with various concentration and high purity, which can be directly applied for enzyme digestion, sequencing, and analysis.

## 1.1  Principle

### 1.1.1  Extraction of plasmid DNA by alkaline lysis method

Plasmid DNA purification includes three basic steps: the growth of the bacterial culture, harvesting, and lysis of the bacteria, and isolation and purification of plasmid DNA.

Bacteria are lysed by treatment with an alkaline solution containing sodium dodecyl sulfate (SDS) and NaOH. While SDS serves to lyses cells and denature proteins, NaOH denatures genomic DNA, plasmid DNA as well as proteins. Then the pH value can be returned to neutral after the supplement of potassium acetate buffer (pH 4.8). Since plasmid DNA is covalently closed circular DNA, it can reanneal accurately and remain in soluble form, while genomic DNA is impossible to renature correctly due to the complete separation of its two strands, which can intertwine to form an insoluble mesh structure. The genomic DNA, unstable macro−molecular RNA and protein−SDS complex can be removed by centrifugation. The remained protein will be removed by phenol/chloroform extraction. The reannealed plasmid DNA from the supernatant is then concentrated by ethanol or isopropanol precipitation.

## 1.1.2　Quantification of plasmid DNA by ultraviolet spectrophotometry

Nucleic acid, nucleotides and their derivatives are aromatic and absorb the ultraviolet (UV) light, whose maximal UV absorption is at 260 nm while the minimum absorption is at 230 nm. Typically, 1 $A_{260}$ is equivalent to a concentration of 50 μg/mL double-stranded DNA or 40 μg/mL RNA.

Protein can also absorb UV light, whose maximal absorption is at 280 nm, and its $A_{260}$ is only 1/10 of that of nucleic acid or lower. For the assessment of the purity of DNA sample, $A_{260}/A_{280}$ ratio is commonly used. $A_{260}/A_{280}$ ratio of RNA is over 2.0, while that of DNA is about 1.8. $A_{260}/A_{280}$ ratio will decrease if the protein content is high, while it will increase if RNA content is high in DNA samples.

$A_{260}/A_{230}$ ratio is also used for the assessment of the purity of nucleic acid, where $A_{260}/A_{230}$ value is commonly in the range of 2.0-2.5. If the ratio is appreciably lower than expected, it may indicate the presence of contaminants, such as phenol and polysaccharides, which can absorb UV light at 230 nm.

## 1.1.3　Identification of plasmid DNA by agarose gel electrophoresis

Agarose gel electrophoresis can effectively identify and purify DNA fragments. At pH 8.0-8.3, the bases in the nucleic acid are almost not dissociable, but all the phosphate groups are dissociable, and thus the nucleic acid will be negatively charged and migrate to the anode in the electric field. According to the charge effect and molecular sieve effect, the nucleic acids with different molecular weights and conformations will have different mobility in gel electrophoresis and thus can be separated efficiently. When the gel is stained by a fluorescent dye, such as ethidium bromide (EB) which can be embedded into the nucleic acids, the bands of nucleic acid can be observed under UV light.

Plasmid DNA has three configurations: covalently closed circular DNA (cccDNA) with super-helix structure, open circular DNA (ocDNA), that is, one strand of cccDNA is broken, and linear DNA, that is, two strands of cccDNA are broken. Those three configurations with the same molecular weights have different mobility in gel electrophoresis, where cccDNA move fastest, followed by linear DNA and ocDNA.

# 1.2　Materials

## 1.2.1　Samples

*E. coli* strain contains plasmid DNA.

## 1.2.2　Reagents and solutions

(1) LB culture medium: dissolve 10 g tryptone, 5 g yeast extract, and 10 g NaCl in 800 mL ddH$_2$O, and adjusted to pH 7.0 by 5 mol/L NaOH. Then add ddH$_2$O to 1 L and autoclave at 15 Ibf/in$^2$ for 20 min, store at 4 ℃.

(2) Ampicillin: 100 mg/mL.

(3) Solution Ⅰ: 50 mmol/L Glucose; 25 mmol/L Tris-HCl (pH 8.0); 10 mmol/L EDTA (pH 8.0).

(4) Solution Ⅱ: 0.2 mol/L NaOH, 1% SDS (prepare before use).

(5) Solution Ⅲ: 5 mol/L potassium acetate : glacial acetic acid : dH$_2$O = 6 : 1.15 : 2.85. The resulting solution is 3 mol/L with respect to potassium and 5 mol/L with respect to acetate.

(6) Phenol/chloroform: mix equal volume of phenol and chloroform, then equilibrium to pH 7.6 by Tris-HCl and store in a brown bottle at 4 ℃.

(7) TE buffer (pH 8.0).

(8) RNaseA (10 mg/mL).

(9) 70% Ethanol and absolute ethyl alcohol.

(10) 6×loading buffer: available commercially.

(11) 10×TAE electrophoretic buffer: 10−folds dilute to 1×TAE (working electrophoretic buffer) prepare before use.

(12) Agarose.

(13) Ethidium bromide (EB, 10 mg/mL): store EB away from the light at room temperature.

(14) DNA molecular weight standard (DNA marker): available commercially.

(15) Plasmid purification kit: available commercially.

## 1.2.3   Special equipment

①Autoclave; ②thermostatic shaker; ③high−speed centrifuge; ④micropipette; ⑤1. 5 mL centrifuge tubes; ⑥electrophoresis tank; ⑦power supply; ⑧gel−making mold; ⑨microwave; ⑩ultraviolet transilluminator; ⑪ultraviolet spectrophotometer.

# 1.3   Methods

## 1.3.1   Procedure

Extraction of plasmid DNA.

(1) Bacterial culture

Inoculate 10 μL *E. coli* which contains plasmid DNA into 10 mL LB medium (containing ampicillin 100 μg/mL), shake at 37 ℃ overnight.

(2) Harvesting of bacteria from the culture

Take 1. 5 mL bacteria solution into 1. 5 mL centrifugal tube, centrifuge at 10,000 r/min for 1 min, discard the supernatant.

(3) Resuspension of bacterial pellet

Add 100 μL of chilled solution I to each tube and resuspend the cell pellet by vortexing or pipetting up and down until no cell clumps remain.

*Note: solution I can disperse bacteria, and chelate metal ions to inactivate DNase which may harm plasmid DNA. Be sure cells are completely resuspended.*

(4) Lysis of bacteria

Add 200 μL solution II and gently invert tube 6−8 times. Place on ice for 2 min.

*Note: do not vortex, as this will result in shearing of genomic DNA. The lysate should appear viscous. Do not allow the lysis reaction to proceed for more than 5 min.*

(5) Neutralization and renaturation of lysate

Add 150 μL chilled solution III, gently invert tube 6−8 times to make lysate homogenous in solution III. Place on ice for 3−5 min.

*Note: after addition of solution III, a fluffy white material forms and the lysate becomes less viscous.*

(6) Clearing of lysate

Centrifuge the bacterial lysate at 12,000 r/min for 10 min.

After centrifuge, two methods can be used to recover plasmid DNA as follow.

## 1.3.2   Protocol A: recover plasmid DNA by ethanol precipitation method

(1) Removing protein by phenol/chloroform extraction

Transfer supernatant containing plasmid DNA (step 6) to a new tube, being careful not to pick up any white precipitate. Add an equal volume of phenol/chloroform solution and shake to mix. Let sit at room temperature for 3 min and centrifuge at 12,000 r/min for 10 min. Then transfer the upper layer solution to another new tube.

(2) Precipitation plasmid DNA by ethanol

Add two volumes of chilled absolute ethanol, mix and let sit at room temperature for 5 min. Centrifugation at 12,000 r/min for 10 min, collect the plasmid in the form of a pellet and discard the supernatant.

(3) Washing of plasmid pellet

Add 1 mL 70% ethanol, gently shaking for several times and then centrifuge at 12,000 r/min for 2 min. Discard the supernatant. This step can remove the salts in plasmid pellet.

(4) Dissolve plasmid DNA

Air–dry the pellet to eliminate all traces of ethanol. Add 30 μL TE buffer (pH 8.0) to dissolve the pellet. Add 2 μL RNase A (10 mg/mL) and incubate at 37 ℃ for 30 min to digest RNA. Store plasmid DNA at −20 ℃.

### 1.3.3   Protocol B Recover plasmid DNA by anion exchanger column (The column and buffer provided in plasmid purification kit)

(1) Equilibration of the column

Put the anion exchanger column on the collection tube. Add 500 μL of elution buffer in the column, then centrifuge at 10,000 r/min for 1 min. Discard the waste solution in the collection tube.

(2) Binding plasmid DNA on the column

Transfer supernatant containing plasmid DNA (step 6) to the column, being careful not to pick up any white precipitate. Place the column at room temperature for 2 min and centrifuge at 10,000 r/min for 1 min. Discard the waste solution in the collection tube.

*Note: by this step, plasmid DNA is believed to adhere to the column via a mechanism called anion exchange, where DNA, a strong anion, binds to the positively charged resin via the salt bridge.*

(3) Washing column twice

Add 600 μL wash buffer in the column, then centrifuge at 10,000 r/min for 1 min. Discard the waste solution in the collection tube. Repeat this step.

*Note: by this step, other material from the lysate, such as proteins, are washed through the column with medium salt buffers.*

(4) Removing wash buffer in the column

Put the empty column on the collection tube, then centrifuge at 10,000 r/min for 2 min.

(5) Collection of plasmid DNA

Put the column on a clean 1.5 mL centrifugal tube, and add 50 μL elution buffer in the center of the column. Stand for 2 min, then centrifuge at 10,000 r/min for 2 min. Store the solution in the centrifuge tube at −20 ℃, which is plasmid DNA solution.

*Note: by this step, purified plasmid DNA is released from the column when a high salt buffer is added disrupting the salt bridge.*

### 1.3.4   Quantification of plasmid DNA by ultraviolet spectrophotometry

(1) Determination of DNA concentration

Take 10 μL plasmid DNA sample. Add 590 μL distilled water (dilute 60 times). Mix well, and then measure the absorbance at 230 nm, 260 nm and 280 nm ($A_{230}$, $A_{260}$, $A_{280}$), respectively, zero the instrument with distilled water.

(2) Calculation of DNA concentration

$$\text{DNA concentration}\left(\frac{\mu g}{\mu L}\right) = \frac{A_{260} \times 50 \times 60}{1,000}$$

## 1.3.5    Identification of plasmid DNA by agarose gel electrophoresis

(1) Prepare 1 % (w/v) agarose gel solution

Add appropriate agarose powder to a measured quantity of 1×TAE buffer in a conical flask. Melt the agarose gel using a microwave until.

(2) Prepare the gel plate

Insert the well-forming comb into the casting tray. When the agarose gel solution described above has cooled to 50-60 ℃, add 2.5 μL EB solution. Mix the gel solution thoroughly by gentle swirling. Then slowly pour the gel solution into the casting tray to avoid air bubbles. The gel could be 3-5 mm thick. When the gel has become solidified, carefully pull the comb straight up and then put the casting tray with gel into the electrophoresis tank. Fill the tank with 1×TAE electrophoresis buffer to cover the gel (1-2 mm above the gel surface).

(3) Load sample

Mix DNA sample and 6×loading buffer. Slowly load 10 μL the mixture into the sample well using a micropipette and add DNA marker into one of the wells. Record sample loading order.

(4) Electrophoresis

Place the cover on the electrophoresis chamber, attach the electrophoresis tank to a power supply with the voltage apply of 60-100 V. As the DNA sample is at the cathode side, DNA sample will migrate toward the anode. Run the gel until bromophenol blue has migrated to a position that is 1 cm away from the leading edge of the gel.

(5) Observe and/or take photos

DNA band in the gel can be observed on an ultraviolet transilluminator, showing a fluorescent band with orange color. Alternatively, take photos with gel imaging system and preserve.

## 1.3.6    Notes

(1) The lysate must be handled gently after addition of solution II and III to prevent shearing of chromosomal DNA.

(2) Carefully control the temperature of the gel solution (not higher than 60 ℃) when making the gel plate, otherwise, it may make the gel-making mold deformation.

(3) Remove the well-forming comb only after the gel is totally solidified. Pull out the comb gently to avoid damage of the wells.

(4) Ethidium bromide (EB) is a strong mutagen, thus, do not contact it with skin during the operation. Always wear disposable gloves when operating EB and be careful not to pollute the experimental environment by EB.

# 1.4    Results and discussion

## 1.4.1    Results

(1) Calculation of DNA concentration.

(2) Electrophoresis.

Plasmid DNA extracted by the alkaline lysis method can be observed in 1-3 bands after separated by agarose gel electrophoresis. For mobility, cccDNA can move fastest, followed by linear DNA and ocDNA. Under normal condition, cccDNA and ocDNA more commonly be observed.

## 1.4.2   Discussion

(1) What is the role of solution Ⅰ, solution Ⅱ, and solution Ⅲ, respectively? What should pay attention to when adding the solution Ⅰ, solution Ⅱ and solution Ⅲ?

(2) Describe the electrophoresis graph of plasmid DNA, and explain the phenomenon and cause.

# 1.5   Troubleshooting

Problems that may be encountered in isolating plasmid DNA and guidance for how to address major difficulties (Table 18.1).

Table 18.1   Plasmid DNA preparation troubleshooting

| Problem | Potential causes | Solutions |
|---|---|---|
| A. Incomplete lysis of bacterial cells | Too many bacterial cells used | Reduce thenumber of bacterial cells or increase the amount of solution Ⅰ, Ⅱ and Ⅲ |
| B. No plasmid DNA or poor plasmid DNA yield | a. Bacterial culture too old | a. Do not incubate cultures for more than 16 h at 37 ℃ under shaking. Pick a single colony from a freshly streaked selective plate |
| | b. Incomplete lysis of bacterial cells | b. Reduce the number of bacterial cells or increase the amount of solution Ⅰ, Ⅱ and Ⅲ |
| | c. Reagents not applied properly | c. Solution Ⅱ and Ⅲ may be turbid under low storage temperatures. Dissolve solution Ⅱ and Ⅲ by warming to 37 ℃ until they change to clear |
| C. Serious RNA contaminates | Too many bacterial cells used, RNase A digestion was insufficient | Reduce the number of bacterial cells or increase the amount of RNase A |
| D. Genomic DNA contaminates | Mixing of bacterial lysate was too vigorous or too much time for lysis of bacteria | The lysate must be handled gently after addition of solution Ⅱ and Ⅲ to prevent shearing of chromosomal DNA. Lyse bacteria less than 5 min |

Problems that may be encountered in agarose gel electrophoresis and guidance for how to address major difficulties (Table 18.2).

Table 18.2   Agarose gel electrophoresis troubleshooting

| Problem | Potential causes | Solutions |
|---|---|---|
| A. DNA sample leaks out the bottom of the well | Bottom of well punctured by micropipette tip | Do not place tip so far down into the well |
| B. DNA bands are smeared | a. Sample degraded | a. Avoid nuclease contamination during the experiment. Prepare a new sample |
| | b. Too much DNA loaded | b. Load less DNA |
| | c. Obsolete electrophoretic buffer | c. Usea new electrophoretic buffer |
| C. DNA bands fuzzy | DNA diffusion due to running gel at very low voltage for a long period of time | Run the gel at a higher voltage for less time |

# Experiment 2　Digestion of Plasmid DNA with Restriction Endonuclease

## 2.1　Principle

Restriction endonucleases (RE) recognize short DNA sequences and cleave double-stranded DNA at specific sites within or adjacent to the recognition sequences. Using a single restriction endonuclease, the circular plasmid DNA can be cleaved into linear DNA with the same size. Therefore, enzyme digestion can be accomplished simply by incubating RE with a certain amount of DNA and the corresponding buffer in the enzyme digestion system for a proper time at the suitable temperature (normally 37 ℃).

Two REs are required for double digestion to form two segments of linear DNA. Double digestion is divided into synchronous digestion and step-by-step digestion, where, synchronous double digestion is more common than the latter, which can economize time and labor. The key is to choose the best buffer for the simultaneous application of the two enzymes. If the buffer can guarantee the high activities of two enzymes, it can be used for double digestion. However, step-by-step digestion can be applied without a suitable buffer for two enzymes at the same time. For step-by-step digestion, the enzyme with low-salt requirement can be applied at first, and then the enzyme with high-salt requirement is supplemented after adjusting the salt concentration to complete double digestion.

## 2.2　Materials

### 2.2.1　Samples

Recombinant plasmid DNA.

### 2.2.2　Reagents and solutions

(1) Restriction endonucleases: available commercially (according to the RE cutting site in recombinant plasmid DNA).

(2) 10×Enzyme digestion buffer: available commercially (accompanied by RE purchase).

### 2.2.3　Special equipment

①Water bath; ②micropipette; ③centrifuge.

## 2.4　Methods

### 2.4.1　Procedure

**Restriction enzyme digestion**

(1) Design enzyme digestion

According to the recombinant plasmid map, select the appropriate restriction enzyme for enzyme digestion.

(2) Prepare the digestion system

The Single digestion is showh in Table 18.3. The Double digestion is shown in figure 18.4.

Table 18.3　Single digestion

| 10×digestion buffer | 2 μL |
| --- | --- |
| Plasmid DNA(1 μg/μL) | 1 μL |
| Restriction endonuclease | 1 μL |
| ddH$_2$O | 16 μL |
| Total | 20 μL |

Table 18.4　Double digestion

| 10×digestion buffer | 2 μL |
| --- | --- |
| Plasmid DNA(1 μg/μL) | 1 μL |
| Restriction endonuclease 1 | 1 μL |
| Restriction endonuclease 2 | 1 μL |
| ddH$_2$O | 15 μL |
| Total | 20 μL |

(3) Perform enzyme digestion

Mix the prepared digestion system and briefly centrifuge. Then incubate the reaction mixture 1 h in the water bath with the corresponding temperature according to the instructions.

(4) Electrophoresis

Stop the reaction and prepare it for agarose gel electrophoresis(See Experiment 1, Chapter 18).

## 2.4.2　Notes

(1) Factors affecting restriction enzyme activity

1) The purity of plasmid DNA: impurities in plasmid DNA, including protein, phenol, chloroform, ethanol, SDS and EDTA, can influence the activity of restriction enzyme. Several methods can improve digestion efficiency, such as purifying plasmid DNA, increasing the enzyme amount, prolonging digestion time and increasing the reaction volume ( >20 μL).

2) Methylation of plasmid DNA: *E. coli* generally has two types of methylase: dam methylase ( modifying A in GATC) and dcm methylase ( modifying C in CCA/TGG). If plasmid DNA is methylated, the digestion efficiency can be reduced. Therefore, the mutant strain with inactivated methylase should be used in genetic engineering.

(3) Buffer: the recognition of restriction enzyme and enzymatic activity develop optimal digestion ability and site specificity under the certain condition such as the ionic strength and pH, thus the specific reaction buffer should be used during the digestion. The restriction enzymes provided commercially can be accompanied with the corresponding optimal buffer to ensure 100% enzymatic activity.

(4) Digestion time and temperature: the optimal reaction temperature of various restriction enzymes was different, mostly with 37 ℃, few with 40–65 ℃. As some restriction enzymes have star activity, the reaction time cannot be too long; otherwise, it may influence the target band.

(5) The concentration of glycerol and enzyme: restriction enzymes are commonly stored in 50% glycerol buffer. The final concentration of glycerol in the digestion system should not exceed 1/10 volume, otherwise, it will inhibit the enzyme activity. The enzyme amount should be controlled to that 1 Unit enzyme can digest

1 μg DNA in 15−20 μL system.

(2) Double digestion

1) If the reaction conditions of the two enzymes are exactly the same (such as temperature and salt concentration), they can be simultaneously added to one tube for the digestion. The current manufacturers can provide a buffer for double digestion or general buffer.

2) If the two enzymes use different temperature, the enzyme with low−temperature requirement should digest at first, then add the second enzyme and increase the temperature to continue the digestion.

3) If the two enzymes require different salt concentrations, the enzyme with low−salt requirement is added firstly and then followed with the high−salt requirement.

## 2.5　Results and discussion

### 2.5.1　Results

(1) Single digestion: one DNA band can be found after complete digestion with the same size of plasmid DNA.

(2) Double digestion: two DNA bands can be found after complete digestion, the sum of the two sizes is identical with that of linear plasmid DNA in single digestion.

### 2.5.2　Discussion

(1) What problems should we pay attention to during the enzyme digestion of plasmid DNA?

(2) What condition is required for the buffer solution in double−enzyme digestion? How to solve this problem without the common buffer solution?

## 2.6　Troubleshooting

Problems that may be encountered in restriction enzyme digestion and guidance for how to address major difficulties (Table 18.5).

Table 18.5　Troubleshooting of restriction enzyme digestion

| Problem | Potential causes | Solutions |
| --- | --- | --- |
| A. Plasmid DNA is not cleaved by restriction enzyme completely | a. The restriction enzyme is inactive | a. The enzymatic activity is determined with the standard substrate |
| | b. Plasmid DNA is impure (containing some inhibiting factors of restriction enzyme, such as protein, phenol, SDS, and EDTA) | b. Purify DNA using column chromatography and precipitate DNA with ethanol |
| | c. The conditions are unsuitable, such as buffer solution and temperature | c. Check whether the reaction condition is optimal |
| | d. The base on the restriction site in plasmid DNA is methylated | d. Replace the same restriction enzyme that is not sensitive to DNA methylation |
| | e. The sequence recognized by the restriction enzyme is not present in plasmid DNA | e. Replace other enzymes to cleave plasmid DNA |

Continue to Table 18.5

| Problem | Potential causes | Solutions |
| --- | --- | --- |
| B. Plasmid DNA is cleaved incompletely | a. The activity of the restriction enzyme is declined | a. Increase the enzyme amount or prolong the digestion time |
| | b. Plasmid DNA is impure (containing some inhibiting factors of restriction enzyme, such as protein, phenol, SDS, and EDTA) | b. Purify DNA using column chromatography and precipitate DNA with ethanol |
| | c. Some DNA solutions stick to the wall of the centrifuge tube | c. Centrifuge few seconds before reaction |
| | d. The reaction conditions are unsuitable | d. Apply the optimal reaction system |
| C. The number of DNA fragments is more than the theoretical value | a. Contaminated by other restriction enzymes | a. Check the digestion result using λDNA as substrate |
| | b. The substrate is contaminated with other DNA | b. Check DNA using electrophoresis, replace other restriction enzymes, and purify DNA fragments |
| | c. The digestion time is too long | c. Shorten the digestion time properly |

# Experiment 3   Isolation, Quantification, and Identification of Eukaryotic Genomic DNA

The extraction, quantitation, and identification of eukaryotic genomic DNA are key steps in genomic analysis, Southern blot, and construction of genomic libraries. Eukaryotic genomic DNA is present in the nucleus so that all nucleated cells can be used to make genomic DNA. The extraction method introduced here is applicable to a large number of mammals DNA extraction.

## 3.1   Principle

The isolation principle of DNA is not only to separate DNA from the complex with proteins, lipids, and sugars but to maintain its integrity. In the extraction solution, SDS serves as a detergent to dissolve plasma and nuclear membrane and denatures proteins. Proteinase K can digest proteins into oligopeptides or amino acids. EDTA can inhibit DNase activity in cells by chelating the $Mg^{2+}$ and prevent DNA from being degraded. Ideally, the cell membrane should be solubilized using detergent at 4 ℃. RNA can be removed by RNase. Protein is removed by shaking the solution gently with phenol/chloroform mixture, which will denature proteins but not nucleic acid.

## 3.2   Materials

### 3.2.1   Reagents and solutions

(1) Homogenizing buffer: 10 mmol/L Tris-HCl (pH 8.0), 25 mmol/L EDTA (pH 8.0), 100 mmol/L NaCl.

(2) 10% SDS.

(3) Proteinase K solution (10 mg/mL).

(4) Pancreatic RNase solution (10 mg/mL).

(5) Saturated phenol : chloroform : isoamyl alcohol = 25 : 24 : 1 (v/v).

(6) Chloroform : isoamyl alcohol = 24 : 1 (v/v).

(7) TE buffer (pH 8.0): 10 mmo/L Tris-HCl, 25 mmol/L EDTA.

(8) 3 mol/L sodium citrate (pH 5.2).

(9) Precool anhydrous ethanol and 70% ethanol.

## 3.2.2   Special equipment

①Refrigerated centrifuge; ②spectrophotometer; ③thermostatic water bath; ④scissors; ⑤glass homogenizer; ⑥shepherd's hook; ⑦vacuum pumping device.

# 3.3   Methods

(1) Tissue homogenate

Weigh 0.1 g fresh animal liver tissue. Rinse 3 times with PBS buffer. Cut with scissors. Place in a glass homogenizer. Add 1 mL homogenization buffer. Then grind the liver to a homogenate on ice.

(2) Lyse cells

Transfer the tissue homogenate to a 1.5 mL centrifuge tube. Add 100 μL 10% SDS and mix. Then incubate at room temperature for 10 min to form a sticky solution.

(3) Digest protein

Add 50 μL proteinase K solution. Gently invert the tube. Incubate in 55 ℃ for 12-18 h. Rotate the tube from time to time.

(4) Digest RNA

Add pancreatic RNase solution to a final concentration of 200 μg/mL. Incubate in 37 ℃ for 1 h.

(5) Centrifuge

Centrifuge at 10,000 r/min for 5 min. Remove the supernatant to a clean centrifuge tube.

(6) Extract

Add an equal volume of phenol/chloroform/isoamyl alcohol. Rotate and mix. Centrifuge at 10,000 r/min for 10 min at 4 ℃. Remove the upper aqueous phase to a fresh centrifuge tube and add an equal volume of chloroform/isoamyl alcohol and mix. Centrifuge at 10,000 r/min for 10 min at 4 ℃.

(7) Purify

Transfer the upper aqueous phase to a clean centrifuge tube. Add 1/10 volume of sodium citrate solution and mix well. Then add 2 times the volume of absolute ethanol and mix. Stand at room temperature for 20-30 min while DNA precipitate forms white flocculent. Pick out with Shepherd's hook and place in a clean centrifuge tube.

(8) Wash

Centrifuge at 10,000 r/min for 10 min. Discard the supernatant and wash the pellet with 70% ethanol. Centrifuge at 12,000 r/min for 2 min. Discard the supernatant and wash the pellet once again. Remove the last trace liquid on the inner wall of the tube by pipette or filter paper. Evaporate the ethanol at room temperature for 15 min.

(9) Collect

Add 200 μL TE buffer to dissolve the pellet. Place the tube on a rocking platform and gently rock the solution overnight at 4 ℃. Store at -20 ℃.

(10) Measure

Determine the concentration of DNA by measuring the absorbance at 260 nm and 280 nm. Identify the quality of the DNA by 0.6% agarose gel electrophoresis.

# 3.4 Results and discussion

(1) Quantitative analysis results

Measure the absorbance of the extracted products at 260 nm and 280 nm ($A_{260}$ and $A_{280}$) and analyze the purity of the extract (see Table 18.6). DNA concentration ($\mu g/mL$) = 50 × $A_{260}$ × dilution factor

Table 18.6　Absorbance of proteins and nucleic acids

| Protein/% | Nucleic acids/% | $A_{260}/A_{280}$ | Protein/% | Nucleic acids/% | $A_{260}/A_{280}$ |
|---|---|---|---|---|---|
| 100 | 0 | 0.57 | 45 | 55 | 1.89 |
| 95 | 5 | 1.06 | 40 | 60 | 1.91 |
| 90 | 10 | 1.32 | 35 | 65 | 1.93 |
| 85 | 15 | 1.48 | 30 | 70 | 1.94 |
| 80 | 20 | 1.59 | 25 | 75 | 1.95 |
| 75 | 25 | 1.67 | 20 | 80 | 1.97 |
| 70 | 30 | 1.73 | 15 | 85 | 1.98 |
| 65 | 35 | 1.78 | 10 | 90 | 1.98 |
| 60 | 40 | 1.81 | 5 | 95 | 1.99 |
| 55 | 45 | 1.84 | 0 | 100 | 2.00 |
| 50 | 50 | 1.87 | | | |

(2) Electrophoresis

Separate the extracted product by 0.6% agarose gel electrophoresis. After EB staining, observe a band near the sample well under UV light (Figure 18.1).

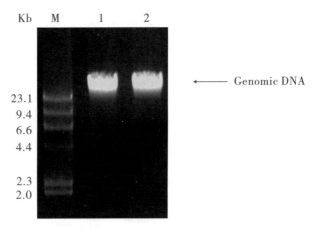

M: DNA Marker; 1,2: sheep genomic DNA.

Figure 18.1　Agarose gel electrophoresis of sheep genomic DNA

# 3.5   Troubleshooting

Problems that may be encountered in isolating genomic DNA and guidance for how to address major difficulties (Table 18.7).

Table 18.7   Troubleshooting of genomic DNA isolation

| Problem | Potential causes | Solutions |
|---|---|---|
| A. After the extraction of phenol/chloroform/isoamyl alcohol, the supernatant is too viscous | Concentrations of DNA is too high | Increase the amount of extraction buffer or reduce the amount of tissue |
| B. Extracted DNA is not easy to dissolve | a. Impure | a. Usephenol/chloroform to extract again and purify DNA precipitate in ethanol |
| | b. Add too little TE buffer | b. Increasethe TE buffer and extend the dissolution time |
| | c. The precipitate is too dry | c. Air-dry not too much |
| C. DNA was diffuse in electrophoresis | Genomic DNA was degraded. The following reason: | |
| | a. The material is not fresh or too much freeze-thaw | a. Take fresh materials. Avoid repeated freezing and thawing |
| | b. Inhibition of endogenous nuclease activity is not well functional | b. Increase the content of the chelating agent in the lysis solution |
| | c. The extraction process is too drastic and DNA was broken | c. Cell lysis should be as gentle |
| | d. Exogenous nuclease contamination | d. All reagents are prepared with sterile water and sterilized at high temperature |
| | e. Repeated freezing and thawing | e. Store the DNA in aliquots to avoid repeated freezing and thawing |
| D. The amount of DNA is too little | a. Sample tissue is not fresh | a. Use fresh materials |
| | b. Insufficient cell lysis caused by uneven mixing of lysate and sample | b. Tissue should be cut into small pieces |
| | c. Inadequate cell lysis caused by the decrease of proteinase K activity | c. Lower precipitation temperature; prolong precipitation time |
| | d. Insufficient cell lysis or incomplete protein degradation due to inadequate warm bath time | d. Extended warm bath time |
| E. Low $A_{260}/A_{280}$ | The sample contains protein or phenol | Extract with phenol/chloroform and purify the DNA with ethanol |
| F. High $A_{260}/A_{280}$ | RNA exist | Test the activity of the RNase used in the experiment |

*Li Zhihong, Peng Fan, Xu Yan*

# Chapter 19

# Isolation, Purification and Identification of RNA

The extraction of RNA is the most essential method used in molecular biology. Quality and integrity of the isolated RNA directly influence the following scientific research. RNA is especially susceptible to degradation due to widespread of RNases in the environment; strong denaturants are therefore generally used in intact RNA isolation to inhibit endogenous RNases. Organic solvent–phenol–chloroform extraction is one of the examples, which is extensively used to isolate RNA. The general steps of RNA purification include four steps: effective destruction; denaturation of nucleoprotein complexes; inactivation of RNase; elimination of protein, carbohydrate, lipids, or DNA.

## Experiment 1　Isolation of Total RNA by TRIzol Reagent and Identification of RNA

## 1. 1　Principle

In the single–step RNA isolation, RNA is separated from DNA in the acidic solution, TRIzol reagent (consisting of guanidinium thiocyanate, sodium acetate, and phenol) and chloroform. The homogenate is allowed to separate into three layers: total RNA will remain in the upper aqueous phase, while DNA and proteins remain in the interphase and lower organic phase, respectively. Recovery of total RNA is then performed by precipitation with isopropanol.

Nucleic acids and proteins absorb specific ultraviolet (UV) at 260 nm and 280 nm, respectively. The "$A_{260}$ unit" is used as a quantitative measure for nucleic acids, while the $A_{260}/A_{280}$ ratio is used to assess the purity of nucleic acids.

In the identification of RNA quality with electrophoresis, intact eukaryotic total RNA emerges as sharp 28S and 18S rRNA bands on a denaturing gel. The approximate 2 : 1 ratio (28S : 18S) of intense is a good indicator of intact RNA.

# 1.2   Materials

## 1.2.1   Reagents and solutions

(1) TRIzol reagent.

(2) Chloroform.

(3) Isopropyl alcohol.

(4) 75% Ethanol (in RNase-free water).

(5) RNase-free water: DEPC (Diethylpyrocarbonate)-treated water.

(6) 10×MOPS running buffer:

0.4 mol/L MOPS, pH 7.0

0.1 mol/L sodium acetate

0.01 mol/L EDTA

(7) Agarose.

(8) Formaldehyde.

(9) Bromophenol blue.

(10) Ethidium bromide.

## 1.2.2   Special equipment

①High-speed microcentrifuge; ②vortex; ③UV (ultraviolet) spectrophotometer; ④electrophoresis equipment; ⑤water bath or heat block; ⑥ultraviolet gel imaging system.

# 1.3   Methods

## 1.3.1   Isolation of total RNA using TRIzol reagent

(1) Homogenization

1) For tissues: add 1 mL TRIzol reagent per 50–100 mg tissue.

For cells grown in monolayer: Add 1 mL TRIzol reagent to a 3.5 diameter dish directly and lyse cells by repetitive pipetting.

For suspension cells: collect cells by centrifugation and lyse cells in 1 mL TRIzol reagent by repetitive pipetting.

*Note: add 1 mL the TRIzol reagent per $(5-10) \times 10^6$ of animal cells. For larger amounts, scale up accordingly. To avoid DNA contamination, sample volume never exceeds 10% of TRIzol volume.*

2) Homogenize the samples on ice using a precooled mortar and pestle.

*Note: it is critical to homogenate the samples as quickly as possible in TRIzol reagent.*

(2) Phase Separation

1) Incubate the homogenate for 5 min at room temperature.

2) Add 0.2 mL chloroform per 1 mL TRIzol reagent (to promote phase separation).

3) Shake tubes vigorously by hand or vortex for 15 s. Incubate the mixture for 2 min at room temperature.

4) Centrifuge the mixture at 12,000 g in a microcentrifuge for 10 min at 4 ℃.

*Note: the mixture separates into three phases: an upper colorless aqueous phase containing RNA, inter-*

*phase containing DNA, and a lower red, the phenol-chloroform phase containing proteins.*

*Note: never disturb the interface.*

(3) RNA precipitation

1) Transfer the upper clear phase to a fresh Eppendorf tube(The volume of the aqueous phase is about 60% of the volume of TRIzol reagent used for homogenization).

2) Add 0.5 mL isopropanol per 1 mL TRIzol reagent used for the initial homogenization.

(optional) If the expected RNA concentration is <10 μg/mL, add a carrier (e.g., glycogen or Glyco Blue) to easily visualize the pellet.

3) Mix vigorously by rapid shaking or vortex.

4) Incubate samples for 10 min at room temperature.

5) Centrifuge at 12,000 g in a microcentrifuge for 10 min at 4 ℃.

6) The white (or blue, if GlycoBlue has been included) RNA pellet often appears as a triangle extending from the side and bottom of the tube slightly upward.

(4) RNA wash

1) Decant the supernatant carefully.

2) Add at least 1 mL 75% ethanol per 1 mL TRIzol reagent used for the initial homogenization. Mix the sample by the vortex.

3) Centrifuge at 10,000 g in a microcentrifuge for 5 min at 4 ℃.

(5) RNA redissolving

1) Decant the supernatant carefully.

2) Briefly, air-dry the RNA pellet for 5-10 min.

*Note: never let the RNA pellet dry completely as its solubility will greatly decrease.*

(6) Dissolve the RNA pellet in 50 μL RNase-free water.

## 1.3.2   Determination of RNA concentration and purity by UV absorption

(1) Pipet 1.0 mL RNase-free water into a quartz cuvette. Place the cuvette in a UV spectrophotometer, read at $A_{325}$, and zero the instrument.

(2) Remove blank cuvette and insert cuvette containing RNA sample (1 μL RNA diluted in 99 μL RNase-free water).

(3) Read the absorbance at 230, 260 and 280 nm.

(4) Use the reading at 260 nm ($A_{260}$) to determine the RNA concentration by the following equation:

$$1 \text{ unit } A_{260} = 40 \text{ μg/mL}$$

(5) Use the ratio of $A_{260}/A_{280}$ and readings at $A_{230}$ and $A_{325}$ to assess the purity of the RNA sample.

## 1.3.3   Agarose gel electrophoresis of RNA

(1) Prepare the gel

1) Dissolve 1 g agarose in 72 mL boiled water, then cool to 60 ℃.

2) Add 10 mL 10×MOPS running buffer, and 18 mL 37% formaldehyde (12.3 mol/L).

*Warning: formaldehyde is toxic through skin contact and inhalation of vapors. Manipulations involving formaldehyde should be performed in a chemical fume hood.*

3) Pour the gel using a comb that will form wells.

4) Assemble the gel in the tank, and add enough 1×MOPS running buffer to cover the gel by a few millimeters.

5) Remove the comb.

(2) Prepare the RNA sample

1) Add 0.5 volume Formaldehyde Load Dye to 1-3 μg RNA.

2) Denature samples by heating at 65-70 ℃ for 10-15 min.

(3) Electrophoresis and photograph

1) Load RNA sample into gel well and electrophorese at 5 V/cm until the bromophenol blue (the faster-migrating dye) has migrated as far as 2/3 the length of the gel.

2) Stain gel in 0.5 mg/mL Ethidium bromide for 2-3 min.

3) Destain gel in RNase-free water for 30 min.

4) Visualize and photograph the gel under a UV transilluminator (305 nm).

# 1.4   Results and discussion

(1) Result of UV absorption

RNA concentration can be determined by a combination of $A_{260}$ reading and the following equation:
1 unit $A_{260}$ = 40 mg/mL

$A_{260}/A_{280}$ ratio of 1.9:2.0 indicates highly purified RNA. This ratio will be lowered by contaminants absorbed at 280 nm.

Absorbance at 230 nm indicates contamination of the sample by phenol, whereas absorbance at 325 nm reflects contamination by particulates and dirty cuvettes.

(2) Results of denaturing gel electrophoresis

Sharp 28S and 18S rRNA bands with 2 : 1 ratio (28S : 18S) of intensity indicates intact total RNA.

# 1.5   Troubleshooting

Problems that may be encountered in UV absorption and gel electrophoresis and guidance for how to address major difficulties (Table 19.1).

Table 19.1   RNA isolation troubleshooting

| Problem | Potential causes | Solutions |
| --- | --- | --- |
| $A_{260}/A_{280}$ ratio<1.8 | A too small volume of TRIzol is used to homogenize the sample | • Scale up the TRIzol volume according to sample |
| | Samples are not stored at RT for 5 min after homogenization | • Store at RT for 5 min after homogenization |
| | The final RNA pellet is not fully dissolved | • Heat to 55-60 ℃ for 10-15 min and repeatedly pipet |
| | Phenol contamination may occur (Absorbanceof phenol is seen at 270 nm) | • Do not attempt to draw off the entire aqueous layer after phase separation |
| | A very large peak is observed from 220-240 nm | • Ethanol precipitation can be performed to remove residual phenol |
| | | • Precipitate and wash again |
| | Residual chloroform | • Do the phase separation after addition of chloroform at 4 ℃ |
| | | • Reprecipitate |
| | $A_{280}$>0.5 | • The reading is out of the linear range, dilute the sample |

Continue to Table 19.1

| Problem | Potential causes | Solutions |
|---|---|---|
| $A_{260}/A_{280}$ ratio>2.0 | RNA degradation occurred | · Avoid washing cells prior to the addition of TRIzol<br>· Immediately process or frozen tissues after removal from the animal<br>· Store RNA after isolation at $-80$ ℃ instead of $-20$ ℃<br>· Cool the samples during homogenization in on-off cycles. |
| Low yield of RNA | Speed vac or lyophilized after the last ethanol precipitation | · Avoid to dry sample completely<br>· Centrifugation speeds should be lower than 12,000 g |
|  | RNA pellet may not be completely solubilized | · Pipet the sample repeatedly<br>· Heat to 50-60 ℃<br>· Perform isopropanol precipitation step with 0.25 volume isopropanol and 0.25 volume high salt solution |
|  | The sample was not fully homogenized | · Complete homogenization |
| Smeared bands or not 2 : 1 ratio | Partial degradation | · Avoid degradation |
| Very low molecular weight smear | Complete degradation | · Avoid degradation |
| No bands visualized using ethidium bromide | <200 ng of RNA was run on the gel | · Scale up sample loading amount<br>· Use SYBR Gold or SYBR Green Ⅱ RNA gel stain instead of ethidium bromide |

*Wang Huaqin*

# Chapter 20

# Polymerase Chain Reaction of Target Genes

Polymerase chain reaction (PCR) is a method for *in vitro* replication of DNA fragments with known sequence conceived in 1983 by Kary Mullis of Cetus Corporation, which facilitates rapid quantification of target genes. PCR is one of the most rapidly developed and popularized molecular biology techniques in recent decades. Since its inception, individual laboratories now have the ability to monitor target gene expression. Moreover, this technique can be combined with other molecular biology methods, which has greatly improved its sensitivity and specificity. Therefore, PCR is widely used in various disciplines of biomedicine.

## Ⅰ. Principles of PCR

(1) Basic principles of PCR

PCR is a process of *in vitro* DNA replication similar to the natural replication process of DNA. Using the catalytic activity of DNA polymerase, a targeted DNA fragment template and two oligonucleotide primers as the starting points for the extension, consecutive cycles of denaturation, annealing, and extension are used to generate a new DNA chain which is complementary to the DNA template. The complementary strand of DNA is synthesized by base pairing and the half preservation replication principle. This technique has become a ubiquitous and powerful tool in gene isolation, sequencing, gene expression regulation, and gene polymorphism research.

(2) The reaction process of PCR

Denaturation: when the temperature of the reaction system is higher than the melting point ($T_m$) of DNA, the high temperature destroys the hydrogen bonds of the DNA double helix. Above the $T_m$, the DNA double helix opens and forms two single linear DNA molecules, forming the template for PCR reactions.

Annealing: when the environment temperature decreases, the two oligonucleotide primers combine respectively on the two sides of the template DNA fragment by base pairing to form a hybridized chain.

Extension: in the presence of the four deoxyribonucleoside triphosphates (dNTPs) and $Mg^{2+}$, under suitable reaction temperatures, DNA polymerase incorporates the appropriate dNTP according to the base paring principle from the 3′ end of the primer, making a new strand of DNA synthesized along the 5′→3′ direction.

The three steps described above compose a PCR cycle and after each PCR cycle, newly synthesized DNA strands serve as templates for the next cycle, therefore the targeted DNA concentration increases at an exponential rate.

# II. Reaction system and conditions of PCR

(1) Reaction system

Despite the numerous variations on the basic theme of PCR, the reaction itself is composed of only a few main components. They are as follows: DNA template, oligonucleotide primers, DNA polymerase, dNTP and reaction buffer.

1) DNA template

The DNA template containing the target gene can be double-stranded or single-stranded, including genomic DNA, plasmid DNA, cDNA or mitochondrial DNA, etc. Although the size of the DNA template is not a key factor, higher molecular weight DNA templates (such as a genomic DNA) will reduce PCR amplification efficiency. One can apply a prior treatment by using restriction endonuclease with rare cleavage sites (such as *SalI* or *NotI*). Amplification efficiency of closed circular DNA is slightly lower than open linear DNA, so it is recommended to linearize the DNA template (such as a plasmid DNA) before performing PCR.

The DNA template used in PCR should be as pure as possible and should not be contaminated with proteases, nucleases, DNA polymerase inhibitors, or proteins that can interact with DNA, etc. RNA contamination can also perturb PCR amplification efficiency, and excess RNA in the DNA template can increase non-specific replication of PCR.

Within a certain range, the output of PCR can be significantly increased with augmentation of DNA template concentration. However, high DNA template concentrations can lead to non-specific replication. In order to ensure the specificity of PCR, one can use ~1 μg genomic DNA or 10-100 ng plasmid DNA for 100 μL PCR reaction.

2) Oligonucleotide primers

Oligonucleotide primers should contain a minimum of 16 nucleotides and 20-24 nucleotides should be targeted to ensure specificity of PCR amplification. When designing PCR primers, one should carefully balance base composition to avoid stacking of purines or pyrimidines. Generally, the appropriate (C+G) base content should be between 40%-60%; lower (C+G) base content can reduce amplification efficiency, while higher (C+G) base content can increase non-specific replication. The optimal primer has a homogeneous distribution of A, T, C, G, and a string of 5 or more purines/pyrimidines should be avoided. In addition, the two primers should not be complementary to avoid the formation of primer dimers. Finally, according to specific requirements, one can insert modifiers on the 5'-end such as restriction cutting sites, mutation sites, biotin or fluorescence, etc.

A concentration of 0.1-1.0 μmol/L is recommended for oligonucleotide primers in PCR reactions. Higher primer concentration can result in mismatches between template DNA and primers and reduce the specificity of the reaction. In addition, high primer concentrations increase the risk of primer dimer formation. In contrast, low primer concentrations reduce the efficiency of PCR amplification.

3) DNA polymerase

There exist two types of commercially available Taq DNA polymerase: one is a natural enzyme extracted from the thermophilic bacteria *Thermus aquaticus*, the other is a cloned enzyme isolated from engineered *E. coli*. For catalyzing 50-100 μL PCR reaction, 1-2.5 U Taq DNA polymerase is generally required.

4) dNTP

dNTP is a mixture of: dATP, dCTP, dGTP and dTTP. The dNTP solution is acidic and it is recommended that a concentrated stock solution is prepared. The pH of the stock solution should be adjusted to approximately 7.0 with 1 mol/L NaOH in order to ensure that the final pH value of the PCR reaction system is not less than 7.1. The concentration of dNTPs used in the PCR reaction is generally 50-200 μmol/L. In the PCR reaction system, the molar concentration of the four dNTP should be identical; imbalance of the dNTP concentration will increase the risk of mismatches by DNA polymerase.

5) Buffer

A standard PCR reaction buffer contains: 10 mmol/L Tris−HCl ( pH 8.3 at RT), 50 mmol/L KCl, and 1.5 mmol/L MgCl$_2$. Mg$^{2+}$ in the reaction buffer is integral as its concentration which is directly related to the activity of DNA polymerase and influences the denaturation temperature of double−stranded DNA. Low Mg$^{2+}$ concentration reduces DNA polymerase activity, thus the PCR production decreases; while high Mg$^{2+}$ concentration perturbs the specificity of PCR reaction. In a standard PCR system, when the dNTP concentration is 200 μmol/L, the Mg$^{2+}$ concentration is generally 1.5 mmol/L.

(2) Reaction conditions

1) Temperature

a. Denaturation: complete denaturation of template DNA is integral to PCR. Generally, a pre−denaturation step is performed at 94−95 ℃ for 3−10 min, and then a denaturation step at 94 ℃ for 30−60 s.

b. Annealing: the annealing temperature is generally 5 ℃ lower than the primers' denaturation temperature. When primers contain 12−25 nucleotides, the annealing temperature can be calculated with the following equation: $T_m = (C + G) \times 4\ ℃ + (A + T) \times 2\ ℃$. The annealing temperate is usually between 40−60 ℃ for 30−45 s. If the (C+G) base content is less than 50%, annealing temperatures should be lower than 55 ℃. Higher annealing temperatures can increase the specificity of PCR.

c. Extension: the extension temperature should correspond with the optimum temperature of Taq DNA polymerase and is generally between 70−75 ℃. The extension duration is dependent on the DNA polymerases extension rate and the length of the targeted DNA fragment. Generally, 1 min is sufficient for a fragment less than 1 kb.

d. Number of amplification cycles: the PCR cycle number is mainly dependent on the quantity of template DNA. Typically, PCR runs consist of 20−30 cycles. Too many PCR cycles can increase non−specific amplification. The optimal number of PCR cycles number should be determined in a preliminary experiment.

e. PCR products and accumulation pattern: amplified PCR products can be divided into two parts: long product fragments and short product fragments. The length of the short product fragment is strictly defined by the two primers. During the first amplification cycle, the original denatured complementary DNA template strands are the only templates in the reaction source. The primers fix on the template and extend in the 5′→ 3′ direction from the primer. Since the extension ends are not the same, the lengths of the amplification products are different. These products with different lengths are called "long product fragments". After entering the second cycle, primers combine with the original template and the newly synthesized chains ( the long product fragments produced in the 1st PCR cycle). The 5′−end of the long product fragment corresponds to the primers' 5′−ends. When the primers fix on the 3′−ends of these fragments, the newly synthesized chains have identical length, and are termed "short product fragments". It is easy to see that the short product fragment increases exponentially ( Figure 20.1), while the longer product is almost negligible as it increases arithmetically, therefore the reaction product does not need to be purified.

2) Other parameters

a. pH: similar to other enzymatic reaction systems, the pH of PCR reaction systems should be stable and adjusted to the optimal pH of DNA polymerase. In the PCR reaction system, 10−50 mmol/L Tris−HCl is generally used to maintain pH between 8.3−8.8. During the extension phase, when the temperature rises to 72 ℃, the pH of the reaction system will be approximately 7.2, the optimal pH of Taq DNA polymerase.

b. The concentration of salts: the salt concentration in the PCR reaction system is an important factor. High salt concentrations favor hybridization of the primer and the DNA template and improve the stability of the hybrids. However, excess ion concentrations inhibit the activity of DNA polymerase. Thus, the salt concentration of the PCR reaction system has to be compromised. The standard PCR buffer contains 50 mmol/L KCl.

c. Reaction additives: in some cases, the additives can enhance the specificity and/or efficiency of PCR such as dimethyl sulfoxide ( DMSO), glycerol, non – ionic detergent, formamide, bovine serum albumin ( BSA), etc. Certain reactions can only be achieved in the presence of these additives. But excess of these reaction additives can inhibit PCR amplification.

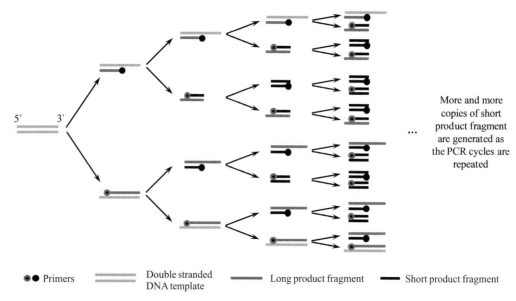

More and more copies of short product fragment are generated as the PCR cycles are repeated

●● Primers     Double stranded DNA template     Long product fragment     Short product fragment

Figure 20. 1    Representation of PCR amplification

During the early stages of the reaction, the DNA fragment of the target gene increases exponentially. After a certain number of PCR cycles, the primers, DNA templates, and DNA polymerase attain a semi–equilibrium, and the amplification process enters a slower production phase, called the plateau phase.

# Experiment 1    PCR Amplification and Identification of Target Genes

## 1.1   Principle

PCR is a process of *in vitro* DNA replication which uses the targeted DNA fragment as a template and two oligonucleotide fragments complementary to the template as primers, under the catalytic activity of DNA polymerase to generate a new DNA chain on the basis of the half preservation replication principle. Repeating this process allows the amplification of the targeted DNA fragment. The main components of a PCR reaction system are as follows: DNA template, oligonucleotide primers, DNA polymerase, dNTP and reaction buffer containing $Mg^{2+}$.

The basic PCR reaction processes include as follows.

( 1 )Denaturation: when the temperature of the reaction system is heated to 94 ℃, the DNA double helix opens and forms two single linear DNA molecules.

( 2 )Annealing: when temperature decreases (5 ℃ lower than $T_m$), the two oligonucleotide primers combine respectively on the two sides of the template DNA fragment.

( 3 )Extension: when re–heating the reaction system to 72 ℃, DNA polymerase synthesizes a new strand of DNA.

(4) The three steps described above compose a PCR cycle and after each PCR cycle, newly synthesized DNA strands serve as templates for the next cycle. As the cycles are repeated, more and more copies are generated and the number of copies of the targeted fragment is increased exponentially.

# 1.2    Materials

## 1.2.1    Reagents and solutions

(1) Primers: 10 μmol/L of up- and downstream primer each.

(2) dNTP: 10 mmol/L (2.5 mmol/L of dATP, dTTP, dGTP and dCTP each).

(3) Taq DNA polymerase: 1 U/μL.

(4) DNA template: 10 ng/μL.

(5) Sterile and deionized water.

(6) 10×PCR buffer (pH 8.3): 100 mmol/L Tris-HCl, 500 mmol/L KCl, 15 mmol/L $MgCl_2$, 1% Triton-100.

(7) 0.5×TBE electrophoresis buffer (pH 8.2) (see Chapter 18, Appendix A).

## 1.2.2    Special equipment

①PCR instrument; ②gel imaging system (UV projection apparatus); ③instrument for horizontal gel electrophoresis.

# 1.3    Methods

## 1.3.1    Procedure

(1) Add the following reagents in a 0.2 mL sterile PCR tube on ice, prepare 50 μL PCR reaction cocktail and gently vortex:

| | |
|---|---|
| DNA template (10 ng/μL) | 1 μL |
| Upstream primer (10 μmol/L) | 1 μL |
| Downstream primer (10 μmol/L) | 1 μL |
| dNTP | 3 μL |
| 10×PCR buffer (pH 8.3) | 5 μL |
| *Taq*DNA polymerase (1U/μL) | 1-2 μL |

Add sterile water to a total reaction volume of 50 μL.

(2) Set the PCR reaction program, slightly centrifuge the PCR tube, place the mixture in the PCR instrument immediately and initiate the amplification reaction. The process is as follow.

1) Pre-denaturation: at 94 ℃ for 5 min.

2) PCR amplification cycle: denaturation at 94 ℃ for 30 s, annealing at 55 ℃ for 30 s, extension at 72 ℃ for 60 s, repeat for 30 cycles.

3) Insulation: keep the PCR system at 72 ℃ for 7 min after the amplification cycle.

4) Stop the reaction: keep the PCR product at 4 ℃ until the following electrophoresis experiment, or at -20 ℃ for long-term storage.

(3) Check and analyze PCR results with agarose gel electrophoresis (see Experiment 1, Chapter 18).

## 1.3.2 Notes

Due to the high sensitivity of PCR amplification, the PCR reaction does not discriminate between targets, the PCR products can be easily contaminated with non−targeted DNA sources and/or unspecific amplification. PCR experiments require the use of a set of positive and negative controls. The positive control is used to test for the presence of inhibitors in the sample or the efficiency of the PCR itself using a pre−dispensed artificial DNA sequence which has moderate amplification capacity and good reproducibility. The negative control consists of a sample without template DNA, to monitor contamination of the system. To evaluate a PCR reaction, the location and size of PCR product bands should be compared with the theoretical value as an important reference index.

# 1.4 Results and discussion

## 1.4.1 Results

PCR amplification products are analyzed using agarose gel electrophoresis (Figure 20.2). To evaluate PCR products, the location and size of product bands must be in accordance with expected values.

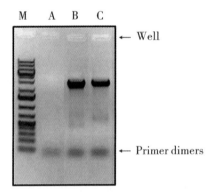

A. negative control; B. positive control; C. a successful PCR amplification experiment.

Figure 20.2　Agarose gel electrophoresis of PCR amplification

## 1.4.2 Discussion

(1) How to design the PCR oligonucleotide primers?

(2) Please describe briefly the effect of the annealing temperature to PCR reaction.

# 1.5 Troubleshooting

Problems that may be encountered in PCR and guidance for how to address major difficulties (Table 20.1).

Table 20.1   Troubleshooting in PCR amplification of the target gene

| Problem | Potential Causes | Solutions |
| --- | --- | --- |
| Non – specific bands: PCR products' bands are not in accordance with expected | The DNA template is contaminated with RNA | Treat the DNA template with RNase |
| | The concentration of DNA template is too high | Reduce the concentration of DNA template |
| | ( C+G) base content in the oligonucleotide primers is too high | Redesign the oligonucleotide primers, reduce the ( C+G) base content |
| | Too many PCR cycles | Reduce the number of PCR cycles |
| Flaky delayed bands or smudged bands | Excess quantity or bad quality of DNA polymerase | Reduce the quantity of DNA polymerase or change the enzyme |
| | Excess concentration of dNTP or $Mg^{2+}$ | Reduce the concentration of dNTP or $Mg^{2+}$ |
| | Low annealing temperature | Increase the annealing temperature |
| | Excess of PCR cycles | Reduce the number of PCR cycles |
| False positive bands | The target gene sequence selected has homology with the non–targeted sequence | Redesign the oligonucleotide primers |
| | Cross–contamination of the target gene or the amplification products | Be cautious to avoid taking wrong samples. Use sterile reagents and materials, do not recycle Eppendorf tubes or pipette tips. Treat reagents and materials with UV before adding samples when necessary in order to destroy non target nucleic acids |

# Experiment 2   Quantitative Reverse–Transcription PCR ( qRT–PCR)

The polymerase chain reaction ( PCR) can be useful to investigate gene expression simply by using RNA transcripts as the initial template. While single–stranded RNA itself cannot be used as a template for PCR, it can easily be converted into double–stranded cDNA using reverse transcriptase. Gene expression can be investigated qualitatively or quantitatively. A quantitative reverse transcription–PCR ( qRT–PCR) methodis described allowing investigation of relative alterations in transcript abundance, revealing how gene expression changes under different circumstance.

## 2.1   Principle

The polymerase chain reaction ( PCR) is one of the most powerful technologies by which a DNA or cDNAtemplate can be copied, or "amplified", many thousand– to million–fold using sequence–specific oligonucleotides, heat–stable DNA polymerase, and thermal cycling.

In quantitative reverse-transcription PCR (often shortened to qRT-PCR), reverse transcriptase and a set of reagents can be used to reverse transcribe RNA templates starting with total RNA, poly(A)+ mRNA or synthetic transcript RNA, then cDNA can be synthesized for subsequent amplification using qPCR. There are three major steps that make up each cycle, including denaturation, annealing, and extension. PCR product is measured at each cycle. By monitoring reactions during the exponential amplification phase of the reaction, the initial quantity of the target can be determined with great precision.

# 2.2 Materials

## 2.2.1 Reagents and solutions

(1) DNase I, amplification grade.

(2) High capacity cDNA reverse transcription kit.

(3) RNase inhibitors.

(4) Power SYBR Green Master Mix.

(5) Oliyo(dT) primers.

(6) Positive control template standards.

(7) DEPC-treated $H_2O$.

Add 1 mL of 0.1% diethylpyrocarbonate (DEPC) per 1 L $ddH_2O$, shake well and incubate overnight with shaking at 37 ℃, autoclave and cool to room temperature before use.

(8) TE buffer.

## 2.2.2 Special equipment

①High-speed microcentrifuge; ②vortex; ③micropipettes; ④nuclease-free microfuge tubes; ⑤real-time thermal cycler.

# 2.3 Methods

## 2.3.1 Reverse transcription reaction (first-strand cDNA synthesis)

(1) Incubate 1 μg total RNA sample at 70 ℃ for 10 min.

(2) Centrifuge briefly, then place immediately on ice.

(3) Prepare a 20 μL reaction by adding the following reagents in the order listed (scale up or down depending on the amount of total RNA) (Table 20.2).

Table 20.2 **Reaction system**

| Component | Volume per 20 μL reaction |
| --- | --- |
| $MgCl_2$, 25 mmol/L | 4 μL |
| Reverse transcription 10× buffer | 2 μL |
| dNTP mixture, 10 mmol/L | 2 μL |
| Oligo (dT) primer | 0.5 μg |

Continue to Table 20. 2

| Component | Volume per 20 μL reaction |
|---|---|
| RNase inhibitor | 20 U |
| RNA sample | 1 μg |
| AMV reverse transcriptase | 15 U |
| Nuclease–free water to a final volume | 20 μL |

(4) ( Optional) Incubate the mixture at room temperature for 10 min if random primers are used.

(5) Incubate the mixture at 42 ℃ for 1 h.

(6) Heat the mixture at 70 ℃ for 15 min.

(7) Incubate the mixture at 4 ℃ for 5 min.

(8) Dilute the cDNA samples to 100 μL with TE buffer.

### 2.3.2   Perform qPCR reactions

(1) Prepare a reaction mix, without template DNA, by combining the reagents in the order listed as follows.

1) 0. 4 μL Forward primer( 10 μmol/L) .

2) 0. 4 μL Reverse primer( 10 μmol/L) .

3) 2 μL template.

4) 10 μL SYBR master mix.

5) Nuclease–Free Water to a final volume of 20 μL.

(2) Carefully add the appropriate volume of the reaction mix to the appropriate wells of the reaction plate. Quickly seal plate after pipetting is complete and protect from light.

(3) Spin down reactions in plate prior to putting a plate in the qPCR machine.

(4) When using an Applied Biosystems 7900HT with SDS 2. 3.

1) Open SDS 2. 3 software, select wells that contain reactions.

2) Add SYBR detector–copy to plate doc.

3) Add dissociation curve to end.

*Note: dissociation curves are useful to detect nonspecific amplification and primer–dimers.*

4) Default parameters are as follows:

①95 ℃ × 2 min.

②95 ℃ × 15.

③60 ℃ × 1 min ( modify this if necessary for each primer set) .

④72 ℃×30 s.

Repeat septs② and ④ for 40 cycles.

⑤60 ℃ × 15 s.

⑥95 ℃ × 15 s

(5) Analyze PCR data using the manufacturer's software and if necessary run reactions on an agarose gel to confirm single correct product.

## 2.4   Results and discussion

There are two primary methods for relative quantitation: relative standard curve and $\Delta\Delta Ct$ ( ddCt) . The efficiency of the assay ( ability to double the amount of DNA/cDNA in every cycle) will determine which method can be used.

(1) Assay efficiency <90%: Relative standard curve

A standard curve is generated by creating a dilution series of sample cDNA and performing PCR. Then, results are plotted with input cDNA concentration on the $x$-axis and Ct on the $y$-axis.

$$\text{relative ratio} = \frac{\text{concentration of target}}{\text{concentration of reference}}$$

(2) Assay efficiency >90%: $\Delta\Delta$Ct (ddCt)

· $\Delta$Ct (sample) = Ct (target gene) − Ct (reference gene)

· $\Delta$Ct (calibrator) = Ct (target gene) − Ct (reference gene)

· $\Delta\Delta$Ct = $\Delta$Ct (sample) − $\Delta$Ct (calibrator)

*Note: The final calculation in $\Delta\Delta$Ct is $2^{-\Delta\Delta Ct}$. The "2" implies a perfect doubling of DNA/cDNA in each cycle of the assay. If assay efficiency is <90%, it is not actually doubling every cycle and the equations for $\Delta\Delta$Ct are no longer valid, requiring either a new start (with bioinformatics) or use of a relative standard curve. It is critical to determine assay efficiency if using $\Delta\Delta$Ct (Table 20.3).*

Table 20.3  Example for calculation

|  | Target gene A | Target gene B | Reference gene |
|---|---|---|---|
| Ct of sample | 20.81 | 20.60 | 15.55 |
| Ct of calibrator | 18.23 | 21.82 | 15.18 |

Expression amount of target gene $A = 2^{(18.23-15.18)-(20.81-15.55)} = 0.2161$

Expression amount of target gene $B = 2^{(21.82-15.18)-(20.60-15.55)} = 3.0105$

This means that compared to the calibrator, the sample exhibits a lower expression of target gene A and a higher expression of target gene B.

## 2.5  Troubleshooting

Problems that may be encountered in qRT-PCR and guidunce for how to address major difficulties (Table 20.4).

Table 20.4  qRT-PCR troubleshooting

| Problem | Potential causes | Solutions |
|---|---|---|
| Formation of primer-dimers | Partial sequence homology exists between the members of the primer pair | · Raise the annealing temperature<br>· Primer concentration can be reduced to 60 nmol/L if necessary<br>· Consider redesigning if necessary<br>· Reverse transcriptase effect on primer-dimers artifact synthesis in RT-PCR. Make sure to thoroughly heat-inactivate the reverse transcription reactions prior to use |
| Poor reaction efficiency | Suboptimal reagent concentrations | · Raise the $Mg^{2+}$ concentration as high as 6 mmol/L<br>· The ideal primer concentration can be anywhere from 100 nmol/L to 600 nmol/L |
| Low expression | Primer $T_m$ values being more than 5 ℃ different from each other<br>Low cDNA input | · The primers are designed to have similar $T_m$ values<br>· The common cDNA input for gene expression analysis is 1-100 ng, but if the gene of interest is of low abundance in the sample, more may be needed |

Continue to Table 20. 4

| Problem | Potential causes | Solutions |
|---|---|---|
| Low expression | Low efficiency of reverse transcription | · Check the type of reverse transcriptase being used<br>· Random primers typically yield more cDNA than oligo( dT) – based methods<br>· Check the reaction components to see if any of these elements can be optimized to improve amplification |
| | The primer designed to the wrong target | · Perform a BLAST$^{TM}$ search against public databases to be sure that the primers only recognize the target of interest<br>· Check sequence databases such as NCBI for variants of the gene of interest |

*Fei Xiaowen, Wang Kai, Wang Huaqin*

# Chapter 21

# Molecular Cloning

## I. Introduction

Molecular cloning, also known as DNA cloning, gene cloning, is a set of experimental methods to construct recombinant DNA molecules by ligating the desired DNA with vector DNA and to direct the chimeric DNA amplified within host organisms. Molecular cloning allows for the production of a large number of target DNA molecules, which can be studied or used to produce the target proteins for other purposes. This technique is based on the manipulation of DNA molecules and the change of cellular genetic character, therefore, as an important part of genetic engineering, molecular cloning is central to many contemporary areas of modern biology and medicine.

The general procedures of molecular cloning experiment include: ①isolation of a target gene from the interested DNA source, and selection of a suitable vector; ②recombination of DNA by ligating the target DNA fragment and vector DNA; ③transfer, to transform the recombinant DNA into a host cell, which is usually a bacterium, or other types of living cell can be used; ④screening and identification of the positive cloning bacteria or cells that contain recombinant DNA; ⑤expression of a cloned gene in the transgenic organism if the target protein product is needed.

## II. Preparation of the target DNA and selection of vector

(1) Preparation of the target DNA

Target gene refers to the gene to be tested or studied. The target DNA can be obtained by chemical synthesis, PCR/RT-PCR, and from genomic DNA library or cDNA library.

If the known DNA fragment is a short sequence, it can be directly synthesized by a DNA synthesizer. The entire genomic DNA of an organism can be digested into DNA fragments with restriction enzymes and packed into vectors to construct different recombinant DNA.

If the sequence of a target DNA on both ends is known, we can design and synthesize a pair of primers, and the target DNA can be obtained by PCR or RT-PCR with genomic DNA or cDNA as templates. This method is simple and specific and can be used to create appropriate restriction sites, add or stop codons on primers, which is currently the most common method to obtain the desired DNA fragment in laboratories.

(2) Selection of vectors

Vectors can serves as a DNA carrier to allow the replication or gene expression of target DNA in the host cell. The common vector has been artificially rebuilt to be cloning vector and expression vector according to the study purpose.

1) cloning vector

Cloning vectors are mainly used to amplify, store or manipulate the target DNA fragments in prokaryotic cells, such as *E. coli*. The artificial vector needs to display several properties for gene cloning, including ①containing an origin (*ori*) of replication; ②having the multiple cloning site (MCS), which contains many unique restriction enzyme cleavage sites; ③ containing selective marker for screening (such as antibiotic resistance marker, etc.); ④the vector should be safety, small size and highly accommodate foreign DNA, and easy to isolate from the host cell. Bacterial plasmids, phages, cosmids, and some artificial chromosomes are commonly used as cloning vectors. Plasmids are small, circular, extrachromosomal DNA molecules found in a variety of bacterias, and the artificial plasmids, such as pBR322, served as the common cloning vectors.

2) Expression vector

Expression vectors allow the exogenous DNA to be stored and expressed to proteins which are coded by the target gene in the host cell. Expression vectors require more features than cloning vectors, including a promoter and terminator for transcription, and translational control sequences, such as a ribosomal−binding site in prokaryotic cell. The target gene should be correctly inserted into an open reading frame (ORF) downstream of a promoter. In addition, tag sequences are usually introduced into some expression vectors to facilitate the detection and purification of the recombinant proteins.

Dependent on the different patterns of gene expression in corresponding host cells, the expression vector is usually divided into two types: the prokaryotic expression vector and the eukaryotic expression vector. According to the expression mode of the target protein, it can be also divided into a non−fusion expression vector, fusion expression vector, and secretory expression vector.

# III. Recombination of target DNA with the vector

In order to insert a gene into a vector to form the recombinant DNA *in vitro*, the ends of the target DNA and vector are usually cleaved using restriction endonucleases (REs) and then ligated by DNA ligase. REs are DNA hydrolases which can specifically recognize and cleave the DNA sequence within the dsDNA. Only Type II RE among three types of REs can cleave DNA at specific recognition sites and can be used as a key tool in recombinant DNA technique. Many REs can recognize the palindrome structure and cut at this sequence, resulting in blunt ends or sticky ends. DNA ligase is another important enzyme which catalyzes phosphodiester bond formation at the ends of DNA fragments and links the recombinant DNA.

(1) Ligation of target DNA and vector after REs treatment

Digestion with REs can result in the formation with a sticky or blunt end. Based on the MCS of the vector and the terminal sequences of the target gene, a single enzyme or double enzyme digestion could be considered in devising cloning strategies. When two REs were used to create the different ends of both target DNA and vector, the target DNA can be placed into the vector in only one orientation, so−called a directional cloning.

If there is not same RE site available for both target gene and plasmid DNA, usually, artificial linkers containing the sticky end of a specific RE cleavage could be added by PCR.

(2) Ligation of PCR products and T vector by TA cloning

At present, TA cloning is the most convenient method for the PCR products cloning. When the target DNA is obtained by PCR, *Taq* DNA polymerase will add a deoxyadenosine (dA) at the 3′ end of PCR products. T vector as a linear DNA with deoxythymidine (dT) at the 3′ end can be purchased from the company, so the ends of T vector and PCR products can directly link with T−A pairing.

# Ⅳ. Transformation of recombinant DNA

Following ligation, the recombinant DNA is introduced into the host cells for replication and gene expression. The host can be prokaryotic cells (such as *E. coli*) or eukaryotic cells (such as yeast, mammalian cell, etc.).

Several methods are available to transfer the recombinant DNA into the host, including transformation, electroporation, microinjection, transfection mediating by calcium−phosphate or liposome, and infection using phage or virus particles.

Transformation is the most common method which introduce recombinant DNA into bacteria and change the genetic traits of the recipient bacteria. The exogenous DNA easily enters into *E. coli* competent cells by means of $CaCl_2$ chemical treatment. In host cells, the foreign gene can be integrated into the chromosome for stable expression, named as stable transfection. It can also be free in the cytosol with relatively short gene expression, which is transient transfection.

# Ⅴ. Screening and identification of positive transformants

Because of the low efficiency of ligation and transformation process, only small part of recombinant vector can be successfully constructed and transferred into the host cell, therefore, a series of methods and techniques are necessary to identify the positive transformant containing the desired gene.

(1) Direct screening of recombinants

Direct screening of recombinants in transformants means the identification of vector and target gene in DNA level.

1) Screening of vector DNA

Common vector usually contains selectable antibiotic resistance markers, such as *Ama^r^*, *Ter^r^*, *Kan^r^*, etc., only bacteria transformed plasmids could grow up as a colony on the antibiotic−containing LB agar plate. In addition, some vectors containing β−galactosidase (*lac Z*) gene provide blue−white screening of the recombinants on X−gal agar plate using the principle of α−factor complementation. At the same time, isolation of plasmid DNA from cultured bacteria colony is also available to identify the successful transformation of the plasmid. However, these methods depend on the vector characters and could not guarantee whether the sequence of target DNA exists correctly in the vector.

2) Identification of target DNA

DNA hybridization (Southern blot) or in situ hybridization of the colony is used to detect the presence of target DNA fragments, but this assay is time−consuming, expensive and it's rarely used.

At present, direct identification of target DNA may be performed by REs analysis, PCR or DNA sequencing. After plasmid DNA is isolated from different colonies, the target DNA fragment could be cleaved with appropriate REs or amplified by PCR using appropriate primers, then the desired DNA fragment would be separated and analyzed in agarose gel electrophoresis or PAGE. Finally, DNA sequencing is the best method to confirm the correct sequence of the inserted gene in the recombinants.

(2) Indirect screening of positive transformants

If the goal of DNA cloning is to express the target gene using expression vector, the expressed products of the target gene could be identified at the protein or mRNA level, named as indirect screening. The methods of Western blot, ELISA and immunoprecipitation (IP) could be utilized to detect the expression of target protein or peptide. Alternatively, the mRNA expression level of the target gene could be analyzed by quantitative PCR.

# VI. Expression and analysis of the cloned gene

In most cases, the practical goal of gene cloning is to express the protein of interest in the host cells for scientific research or clinic applications. The expression system is constituted of an expression vector with appropriate transcriptional and translational elements and the host cell which is suitable for foreign gene expression. According to the type of host cells, expression systems can be classified into prokaryotic expression system and eukaryotic expression system. Expression of the eukaryotic gene could be carried out in prokaryotic expression system or eukaryotic expression system.

( 1 ) Expression system

*E. coli* expression system is one typical prokaryotic gene expression system that widely used in target gene expression, because *E. coli* is the most popular host due to its easy culture, fast proliferation, and inexpensive in large quantities under the laboratory condition. The target products are expressed at high levels in *E. coli*. induced by IPTG. However, the expression of eukaryotic proteins in *E. coli* also has some shortcomings: ①*E. coli* lacks post−transcription processes, the only cDNA of a eukaryotic gene can be expressed. ②*E. coli* lacks post−translation modifications that are necessary for eukaryotic genes, so eukaryotic proteins may not be correctly folded and modified in *E. coli*. ③The expressed protein usually forms insoluble inclusion body with an inactive structure, furthermore, eukaryotic proteins may be recognized and degraded by protease in *E. coli*.

Eukaryotic expression system( yeast, mammalian cell system, etc. ) equipped with proper transcriptional and translational elements has been designed to obtain the bioactive protein with corrected folding and modification, although eukaryotic cells require more strict culture condition, and the process of gene expression is time−consuming, low yield and complicated manipulation.

( 2 ) Protein tags

Whether prokaryotic expression or eukaryotic expression, expression vectors often contain a tag sequence just upstream of the MCS that facilitate the expression, identification, and purification of protein products. The commonly used tags include His−tag, GST−tag, HA−tag, Flag−tag, etc. His−tag as a six or more histidine peptide, normally does not affect the conformation or folding of fusion target proteins. The specific binding with some transition metal ion ( e. g. , $Ni^{2+}$ or $Co^{2+}$ ) could be employed to purify recombinant His−tag protein by affinity chromatography. Another frequently used protein tag, Glutathione S−transferase ( GST) is often co−expressed with the target gene and formed soluble recombinant proteins. GST tagged fusion protein can be purified by affinity binding with Glutathione ( Glu) agarose beads because GSH is the specific substrate of GST. However, GST is often needed to remove after purification because of its large molecular weight ( 26 kDa). Green fluorescent protein ( GFP) is also one common protein tag composed of 238 amino acids, which excites obvious green fluorescence when exposed to light at $\lambda_{498 \text{ nm}}$. GFP with the fusion protein or independent protein tag can be checked by fluorescence−based technologies. Normally tag−fusion proteins can be easily detected or confirmed the localization by Western blotting or immunohistochemical assay.

# Experiment 1   TA Cloning of PCR Products

## 1.1   Principle

PCR products amplified by *Taq* DNA polymerase will be added a deoxyadenosine (dA) at the 3′ end, while the linearized T vector has deoxythymidine (dT) at the 3′ end, so PCR products can be directly linked to T vector with T–A pairing by DNA ligase. The pGEM–T (easy) vector from Promega and the pMD18–T vector from Takara are the most widely used T vectors in most laboratories.

The pMD18–T vector is constructed by adding the *Eco*R V recognition sequence in MCS of pUC18 vector (Figure 21.1). The T vector will be linearized with 3′ protruding T ends of both sides after cleaved by *Eco*R V. The pMD18–T vector also maintains the selective markers, including an ampicillin resistant (*Amp*$^r$) gene and a β–galactosidase (*lac Z*) gene, which could be used for Amp plate screening or blue–white screening.

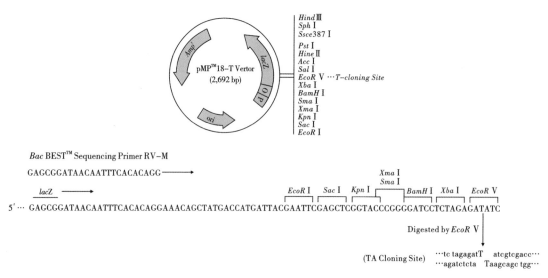

Figure 21.1   **The map of pMD18–T vector**

After ligation, the recombinant DNA is transferred into *E. coli* for cloning. The competent *E. coli* cells will be prepared by the treatment with cold (0–4 ℃) CaCl$_2$, which temporarily change the membrane permeability, and DNA can form hydroxyl–calcium phosphate compound which adheres on the surface of bacteria. With transient heat shock treatment at 42 ℃, DNA molecules can easily enter into *E. coli*, then the competent cells are revived and cultured in selective LB solid medium (containing proper antibiotics, such as Amp), and the positive transformant transferred by the plasmid will grow up into the colony on the selective plate. T–vector can also be selected by the α–complementary method, namely blue–white screening.

# 1.2   Materials

## 1.2.1   Samples

(1) PCR products: amplified by *Taq* DNA polymerase.

(2) pMD18-T vector.

(3) Recipient cells (*E. coli*).

## 1.2.2   Reagents and solutions

(1) 10×ligase buffer.

(2) T4 DNA ligase (2 U/μL).

(3) Double distilled water (ddH$_2$O).

(4) Ampicillin (Amp): dissolve 100 mg/mL Amp in ddH$_2$O, store at −20 ℃, the working concentration is 50-100 μg/mL.

(5) LB liquid medium(1 L): dissolve 10.0 g Tryptone, 5.0 g yeast extract, and 10.0 g NaCl completely in ddH$_2$O, and adjusted to pH 7.0 by 5 mol/L NaOH. Then add ddH$_2$O to 1 L and autoclave at 15 Ibf/in$^2$ for 20 min.

(6) LB solid medium: add 1.5 g agar powder into 1 L LB liquid medium, autoclave at 15 Ibf/in$^2$ for 20 min, cool to about 60 ℃. Add Amp to 50-100 μg/mL and mix well, then pour the medium on culture dishes.

(7) 0.1 mol/L CaCl$_2$ solution: autoclave, store at 4 ℃.

(8) Single colony or frozen species.

(9) PCR product purification kit.

## 1.2.3   Special equipment

①Thermostatic water bath; ②thermostat shaker.

# 1.3   Methods

## 1.3.1   Preparation of competent cells

(1) Bacteria mini-preparation

Pick a single bacteria colony from LB medium plate and transfer to 3 mL LB liquid medium, incubate in thermostat shaker at 37 ℃(200-300 r/min) overnight.

(2) Exponential amplification of cultured bacteria

The next day, transfer 1.0 mL overnight cultures into the flask with fresh 100 mL LB medium, shake vigorously at 37 ℃(−300 r/min) for 2-3 h, until the $A_{600}$ value reaches 0.3-0.4, remove the flask immediately to ice bath for 10-15 min.

(3) Bacteria collection

Transfer the bacteria to a pre-chilled 50 mL sterile centrifuge tube and centrifuge at 4,000 g, 4 ℃ for 10 min. Discard the supernatant and place the tube upside down on the filter paper for 1 min to drain the remaining medium.

From this step, the operation should not be polluted by other bacteria.

(4) $CaCl_2$ treatment

Resuspend bacterial gently with pre-chilled 10 mL 0.1 mol/L $CaCl_2$, place in ice bath for 30 min. Centrifuge at 4 ℃, 4,000 g for 10 min. Discard the supernatant and place it on the filter paper for 1 min.

Resuspend bacterial gently with 4 mL pre-cooled 0.1 mol/L $CaCl_2$. Immediately dispense every 0.1 – 0.2 mL cells into an autoclaved Ep tube and store in a refrigerator at 4 ℃ for 12–24 h, ready for transformation.

*Note: competent cells can be stored at –70 ℃ with 30% glycerol for several months.*

### 1.3.2  PCR product purification

(1) Obtain target gene by PCR (see Experiment 1, Chapter 20)

(2) According to the instruction of purification kit, add the corresponding volume of binding buffer to each PCR tube (different kit, different ratio of volume), mix sample well.

(3) Insert a high pure filter tube into one collection tube. Transfer the mixture to the filter tube, centrifuge at 10,000 g for 1 min.

(4) Discard the flow-through solution in the collection tube. Add 500 μL wash buffer to the filter tube, centrifuge at 10,000 g for 1 min.

(5) Discard the flow-through solution in the collection tube. Add 200 μL wash buffer to the filter tube, centrifuge at 10,000 g for 1 min.

(6) Discard the flow-through solution in the collection tube, centrifuge filter tube at 10,000 g for 1 min.

(7) Transfer the filter tube to a clean 1.5 mL Ep tube. Add 50 μL Elution buffer to the filter tube, stand at room temperature for 5 min, centrifuge at 10,000 g for 1 min, then the purified PCR product is in the Ep tube.

### 1.3.3  The ligation reaction between the target gene and the T vector

Ligate the target gene from PCR product with the T vector.

Establish the following reaction in a sterile Ep tube(Table 21.1).

Table 21.1  The ligation reaction system

| Components volume ( μL) | Volume/μL |
|---|---|
| PCR product (0.1~0.3 pmol/L) | 1 |
| pMD18-T vector (50 ng/μL) | 1 |
| 10 × ligase buffer | 2 |
| T4 DNA ligase (2 U/μL) | 1 |
| ddH$_2$O | 15 |
| Total | 20 |

Mix well, incubate the reaction tube in water bath at 16 ℃ for 12–16 h or overnight, and the target gene will be inserted into the vector. Then the reaction solution containing recombinant DNA is collected for transformation.

### 1.3.4  Transformation

(1) Take 100 μL competent cell solution into a sterile Ep tube, add 10 μL the ligation product, gently rotate and mix the contents, then place the tube on ice for 30 min.

(2) Heat shock: place the tube in the 42 ℃ water bath for 90 s, do not shake the tube, then immediate-

ly place on ice for 1−2 min.

(3) Bacterial recovery: add 400 μL LB medium without antibiotics into the tube, gently shake at 37 ℃ (100−150 r/min) for 45−60 min. Bacteria will be reanimated and express the $Amp^r$ gene of the plasmid.

### 1.3.5    Recombinant DNA screening

(1) Plates screening: spread 0. 1 mL bacterial solution on LB solid medium containing 100 μg/mL Amp, put the plates at 37 ℃ for 15−20 min, then incubate the plates upside down at 37 ℃ for 12−16 h, not more than 20 h.

(2) Recombinant DNA identification: pick several bacteria colonies grown up on plate and culture with Amp−containing LB liquid media, then isolate the plasmid DNA and identify whether they contain recombinant DNA using PCR or RE digestion reaction (see Experiment 2, Chapter 18), or DNA sequencing.

### 1.3.6    Notes

(1) The ratio of PCR product and vector DNA should be (2−10) ∶ 1. DNA purity has a strong effect on the efficiency of ligation, so PCR product should be purified using the purification kit.

(2) For ligation, DNA cyclization will be inhibited if the temperature exceeds 26 ℃.

(3) For transformation, a positive control (using a known amount of supercoiled plasmid DNA) should be set up to estimate transformation efficiency; a negative control (without any plasmid, only competent cells in the tube) should be set up to eliminate possible contamination and easy to pinpoint the reason.

(4) The best bacterial growth density is about $5×10^7$/mL (measured $A_{600}$ = 0. 3−0. 4), whether the density is too high or too low, the transformation rate will be reduced.

(5) The plasmids used for transformation are mainly covalently closed loop DNA, and the transformation rate is proportional to the concentration of exogenous DNA.

(6) Prevent the contamination of bacteria and other foreign DNA, otherwise, it will affect the transformation rate or turn into hybrid DNA.

(7) For screening, the plating density of the transformant should be low, and culture time does not exceed 20 h. Overload density or prolonged incubation time can lead to $Amp^r$ satellite colonies. There are not linearly proportional between the increased anti−Amp colonies and the increased bacteria, probably because cells killed by the antibiotics release growth−inhibiting substances.

## 1.4    Results and discussion

### 1.4.1    Result

A large number of single colony growth can be seen after incubation, those are positive colonies, which indicate that plasmid transformation is successful. But it does not mean there is recombinant DNA in the positive colony. The plasmid in the positive colony should be identified by PCR or restriction enzyme analysis.

### 1.4.2    Discussion

(1) Why should set up a positive control and negative control in the transformation experiment?

(2) In order to obtain competent cells of *E. coli* with high activity, what details need to be noted in the preparation process?

(3) Why culture time should not be too long while screening $Amp^r$ host cell of *E. coli*?

# 1.5   Troubleshooting

Problems that may be encountered in TA cloning and guidance for major difficulties( Table 21. 2) .

Table 21.2   Troubleshooting in TA cloning

| Problem | Potential causes | Solutions |
|---|---|---|
| · PCR product | · There are ligation – inhibiting components in PCR product solution | · PCR product should be purified before ligation |
| | · The PCR product does not contain a 3′–A terminus | · Using suitable Taq DNA polymerase |
| · T vector | · T–terminal deletion in T vector | · Exonuclease contamination should be avoided which cause T–terminal degradation |
| · T4 DNA ligase and buffer | · The activity of T4 DNA ligase and buffer are decreased | · Pay attention to the preservation of commercial ligase and the buffer dilution ratio |
| · Ligation time | · Ligation time is too short | · The ligation time should be set for overnight in 16 ℃ water bath |
| · There is not any positive colony in the plate | · Recombinants are too large to enter the bacteria | · Choosing another suitable vector |
| | · Competent cells are in poor condition or have died | · Using new fresh competent cells |
| | · During bacteria recovery, LB liquid medium contains Amp, so the bacteria already died | · Using LB liquid medium without antibiotic |
| | · Heat shock time is too long | · Strictly control of time |
| · Low transformation efficiency | · Low ligation efficiency or low activity of competent cells | · Optimizing the condition of ligation, or using new and fresh competent cells |
| · Too many colonies | · Too much bacteria is plated | · An appropriate amount of bacteria should be plated, and smear uniformly |
| · There have colonies in the negative control | · Competent cells are contaminated by antibiotic–resistant strains | · Using new and fresh competent cells |
| | · Selective plate failure because antibiotic does not work | · Preparing fresh antibiotic and controling the concentration |

# Experiment 2   Subcloning of Target Genes

# 2.1   Principle

Molecular subcloning is a technique used to transfer the target gene from the parent vector to a destination vector. The aim of subcloning is to analyze and modify the target gene DNA or induces the target gene expression for purification and functional or structural analysis. Because of the simple structure of vector, fast growth of *E. coli*, and low-cost culture expense, the plasmid is normally used to amplify the target gene. However, the eukaryotic protein should be modified after synthesis, which owns natural structure and activity, therefore, in this subcloning experiment, the eukaryotic target gene constructed with cloning vector will be subcloned into eukaryotic expression vector, such as pGEX-2T, for protein expression.

During the process of subcloning, the target gene could be obtained from recombinant DNA by PCR or digestion reaction with the corresponding REs. At first, the growing bacterial colonies containing recombinant DNA are amplified, harvested and extracted the recombinant DNA. Subsequently, both of purified recombinant DNA and pGEX-2T plasmid are digested by REs and the DNA fragments are separated by low melting point agarose gel electrophoresis. The target DNA fragment and linear pGEX-2T vector can be recycled from agarose gel by phenol extraction or DNA adsorption kit since this agarose gel has a low melting point ( <65 ℃).

Depending on the desired DNA and vector, one or two of the appropriate REs cleavages can be used to generate cohesive ends or blunt ends for ligation, if asymmetric sticky ends are digested with two different REs, DNA ligation would be easy to perform as directional cloning. Finally, the new recombinant DNA is transformed into appropriate host cells, and identified by different screening methods.

## 2.2   Materials

### 2.2.1   Samples

(1) *E. coli* containing recombinant DNA with the cloning vector.
(2) Expression vector pGEX-2T.

### 2.2.2   Reagents and solutions

(1) 10 ×RE buffer.
(2) BamH Ⅰ (10 U/μL).
(3) EcoR Ⅰ (10 U/μL).
(4) DNase-free ddH$_2$O.
(5) EDTA. Na$_2$(pH 8.0):100 mmol/L.
(6) Low melting point agarose.
(7) Ethidium bromide (EB).
(8) TE buffer (10 mmol/L Tris · Cl, 1 mmol/L EDTA, pH 8.0).
(9) Balance phenol (saturated 0.1 mol/L Tris · Cl, pH 8.0).
(10) Chloroform/isoamyl alcohol (24 : 1).
(11) 10 mol/L ammonium acetate (NH$_4$Ac).
(12) Absolute ethanol and 70% ethanol.
(13) 10 × ligase buffer.
(14) T4 DNA ligase (2 U/μL).
(15) Competent cell(see Experiment 1, Chapter 21).
(16) LB liquid medium(see Experiment 1, Chapter 21).
(17) LB solid medium(see Experiment 1, Chapter 21).
(18) Ampicillin: dissolve 100 mg/mL Ampicillin in ddH$_2$O, store at −20 ℃, the working concentration is 50-100 μg/mL.

### 2.2.3   Special equipment

①Water bath; ②micro samplers; ③electrophoresis tank; ④vortex mixer; ⑤gel imaging system.

# 2.3 Methods

## 2.3.1 Isolation of the target DNA and expression vector

(1) Prepare selected expression vector (such as pGEX-2T vector) and recombinant DNA constructed by cloning vector (refer to Experiment 1, Chapter 18).

(2) Digest vector and recombinant DNA with REs in a sterilized Ep tube, respectively.

Table 21.3  RE digestion system

| Composition | Volume/μL |
| --- | --- |
| Target gene or vector DNA (1/μg) | 10 |
| *Bam*H I (10 U/μL) | 1 |
| *Eco*R I (10 U/μL) | 1 |
| 10 ×RE buffer | 2 |
| ddH$_2$O (DNase free) | 6 |
| Total volume | 20 |

(3) Separate the digested products through 1% low melting temperature agarose gel electrophoresis (agarose gel electrophoresis refers to Experiment 1, Chapter 18).

(4) Stain the gel with ethidium bromide (EB) and cut out the desired DNA band with a scalpel under the gel imaging system, transfer the gel slice to an Ep tube.

(5) Add 2-3 gel slice volumes of TE buffer, and incubate at 70 ℃ for 10 min to melt the gel (agarose percentage should be reduced to 0.4% or less).

(6) Cold to room temperature, add an equal volume of equilibrium phenol, mix vigorously and centrifuge at 4,000 g for 10 min. The interface of the organic phase and the aqueous phase is white low melting point agarose powder.

(7) Collect the water phase to a new Ep tube, add an equal volume of chloroform/isoamyl alcohol to extract DNA fragments.

(8) Transfer the upper water phase to a new Ep tube, add 0.2 volume of 10 mol/L NH$_4$Ac and 2 volumes of ethanol, mix sufficiently and set at room temperature for 10 min to precipitate DNA.

(9) Centrifuge at 4 ℃, 12,000 g for 10 min. Discard the supernatant, and rinse the DNA pellet with 70% ethanol to wash the DNA.

(10) Centrifuge at 4 ℃, 12,000 g for 5 min, discard the supernatant, and dry pellet.

(11) Resuspend the target DNA fragments in an appropriate amount of ddH$_2$O.

## 2.3.2 Ligation of target DNA and expression vector

(1) Add the target DNA fragment, and linear pGEX-2T vector in a new Ep tube as follows (Table 21.4).

Table 21.4   The ligation system

| Composition | Content |
| --- | --- |
| Purified target gene fragments | 1 μL(0.4 μg) |
| Linear pGEX−2 T vector | 1 μL(0.1 μg) |
| 10 × Ligase buffer | 2 μL |
| T$_4$ DNA ligase (2 U/μL) | 1 μL |
| ddH$_2$O ( without DNase) | 5 μL |
| Total volume | 10 μL |

Mix well, incubate at 16 ℃ for 12−16 h.

(2) Transformation and plates screening: take 10 μL product and transform it into competent cells, and the positive colonies on Amp−containing LB solid plate ( refer to Experiment 1, Chapter 21).

(3) Recombinant DNA identification( refer to Experiment 1, Chapter 21).

(4) Identification of the target protein: if target DNA in the cloning vector will be subcloned into the expression vector, such as pGEX−2T, for protein expression ( refer to Experiment 3, Chapter 21). The expression level and specificity of target protein should be identified at the protein level by SDS−PAGE ( refer to Experiment 2, Chapter 8) or by Western blot ( refer to Experiment 3, Chapter 8).

## 2.3.3   Notes

(1) Foreign DNA contamination should be avoided.

(2) After the electrophoresis, observe the specific location of the target fragment in the UV transilluminator and cut out the gel band with a clean scalpel. A new plastic wrap should be placed under the gel to avoid contamination. Do not use short−wave UV light (254 nm), which causes DNA breakage or TT dimers formation. The former will make the ligation and transformation failed, and the latter may cause genetic mutation. In order to improve the recovery efficiency of DNA fragments, try to remove excess gel.

(3) The recovered DNA fragments are with high purity, which can be used directly for DNA sequencing and cloning experiments, as well as various enzymatic reactions.

(4) The experimental operation should be gentle, especially when dealing with larger fragments, the damage of DNA by mechanical shearing action should be prevented.

(5) Most of the REs are stored in 50% glycerol buffer at −20 ℃. In the enzymolysis reaction, more than 5% of the glycerine concentration inhibits the RE activity, so the amount of the enzyme should be limited to one−tenth of the total volume.

(6) Excessive endonuclease (2−5 times) can shorten the reaction time, but the amount of enzyme can not be too much, because many of the endonuclease itself will decrease the specificity of the recognition sequence.

(7) The molar ratio of target DNA and vector DNA is usually (3−5) ∶ 1 for ligation.

(8) The 5′−P of the vector is removed to inhibit plasmid self−ligation and cyclization.

(9) To terminate the digestion reaction, EDTA can be added to a final concentration of 10 mmol/L.

# 2.4   Results and discussion

## 2.4.1   Results

After REs digestion and electrophoresis, the target gene and linear vector should be separated. Two predicted DNA fragments with a large size and a small size appear individually in the gel.

## 2.4.2 Discussion

(1) When the target DNA fragment is obtained by gel recycling, what aspects should be paid attention to in order to ensure DNA integrity?

(2) To express plenty of a human protein, how to improve the efficiency of protein expression?

## 2.5 Troubleshooting

Problems that may be encountered in subcloning and guidance for how to address major difficulties (Table 21.5).

Table 21.5 Troubleshooting in subcloning

| Problem | Potential causes | Solutions |
|---|---|---|
| Low recovery | a. The gel melting is uncompleted | a. Cut the gel into small pieces as much as possible and make sure the gel is thoroughly melted |
| | b. The amount of ethanol in the cleaning liquid is insufficient | b. Add ethanol in proportion |
| The recycled product is not clear strip by electrophoresis | a. DNA molecular is broken | a. Mixing and other operations should be as gentle as possible |
| | b. Contamination of other DNA | b. Using the newly prepared agarose gel and electrophoresis buffer |
| | c. DNA degradation | c. Avoiding DNA degradation |
| No expected restriction fragments | a. Endonuclease activity decreased | a. Check the endonuclease activity, storage conditions |
| | b. Reaction conditions are not suitable | b. Check the reaction buffer concentration |
| | c. DNA purity is not enough | c. Repurification by ethanol precipitation |
| | d. Restriction sites were modified | d. Choose a suitable isoschizomer |
| | e. DNA sticky end annealing | e. DNA electrophoresis stored at 65 ℃ for 10 min |
| More DNA fragments than expected | a. Endonuclease contamination | a. Check the purity of endonucleases |
| | b. Contain other impurities DNA | b. Repurification of DNA |

# Experiment 3  Induced Expression and Purification of GST-tagged Fusion Protein

## 3.1  Principle

To obtain a great number of recombinant proteins at high purity is a prerequisite for downstream scientific or medical applications. In *E. coli* expression system, the expression of target genes can be induced by IPTG. Protein tags are fused with the target gene and co-expressed, which facilitating the purification and identification of target proteins.

The recombinant plasmid pGEX-2T contain GST tag, which has a molecular weight of 26 kDa, the target gene can be inserted into the vector and co-expressed with GST induced by IPTG in *E. coli* BL21 strain. The fusion protein will be purified by affinity chromatography using Glutathione-agarose beads. As a special substrate of GST, reduced GSH that immobilized in solid support can be used for binding GST-tagged protein by the enzyme-substrate reaction. Then GST-tagged protein can be washed and collected by competitive elution with GSH.

# 3.2　Materials

## 3.2.1　Samples

(1) Recombinant DNA constructed by pGEX-2T plasmid.

(2) *E. coli* BL21 competent cells.

## 3.2.2　Reagents and solutions

(1) LB medium (See Experiment 1, Chapter 21).

(2) IPTG (stock at 1 mol/L, -20 ℃).

(3) 10×PBS buffer, dilute in 1×PBS buffer before use (1 L) (pH 7.4) (NaCl 80.0 g, KCl 2.0 g, $Na_2HPO_4 \cdot 12H_2O$ 35.8 g, $KH_2PO_4$ 2.4 g).

(4) Ampicillin (100 mg/mL).

(5) Glu-agarose beads.

(6) Lysis buffer (100 mL) (prepare as described in Table 25.6).

Table 25.6　Preparation of lysis buffer

| Component | Final concentration | Stock solution | Amount |
|---|---|---|---|
| PBS buffer (pH 7.2) | | | 90 mL |
| DTT | 1 mmol/L | 1 mol/L | 0.1 mL |
| Benzonase | 3 U/mL for bacteria culture | | 600 U |
| Triton X-100 | 1% | | 1.0 mL |
| EDTA | 1 mmol/L | 0.5 mol/L | 0.2 mL |
| Lysozyme | 1 mg/mL | 100 mg/mL | 1.0 mL |
| PMSF | 1 mmol/L | 1 mol/L | 0.1 mL |
| Add ddH₂O to 100 mL, prepare fresh before use | | | |

Wash buffer (100 mL) (prepare as described in Table 25.7).

Table 25.7　Preparation of wash buffer

| Component | Final concentration | Stock solution | Amount |
|---|---|---|---|
| PBS buffer (pH 7.2) | | | 90 mL |
| DTT | 1 mmol/L | 1 mol/L | 0.1 mL |
| EDTA | 1 mmol/L | 0.5 mol/L | 0.2 mL |
| Add ddH₂O to 100 mL, prepare fresh before use | | | |

Elution Buffer (100 mL) (prepare as described in Table 25.8).

Table 25.8  **Preparation of elution buffer**

| Component | Final concentration | Stock solution | Amount |
|---|---|---|---|
| Tris–HCl (pH 8.0) | 50 mmol/L | 1 mol/L | 5.0 mL |
| NaCl | 100 mmol/L | 1 mol/L | 10.0 mL |
| DTT | 1 mmol/L | 1 mol/L | 0.1 mL |
| Triton X–100 | 0.1% | | 0.1 mL |
| GSH | 50 mmol/L | 1 mol/L | 5.0 mL |
| Add ddH$_2$O to 100 mL, prepare fresh before use | | | |

All reagents for SDS–PAGE and western blotting (refer to Experiment 2–3, Chapter 8)

### 3.2.3  Special equipment

①Laminar flow cabinet; ②temperature controlled shaker; ③mini–protein electrophoresis equipment with power supply.

## 3.3  Methods

### 3.3.1  Induced expression of the fusion protein

(1) The recombinant DNA constructed by pGEX–2T plasmid is transformed into *E. coli* BL21 competent cells. The bacterial colony is picked out and transferred into 3 mL Amp–containing LB medium, and incubated overnight in a shaker at 37 ℃.

(2) Transfer 0.2 mL overnight cultures into 20 mL fresh Amp–containing LB medium in a flask, grow at 37 ℃ with shaking vigorously (–300 r/min) for 3–4 h until the OD$_{600}$ value reaching 0.3–0.5.

(3) Collect 1 mL cultured medium, centifuge at 10,000 g for 1 min, discard the supernatant, the bacteria pellet as a control sample before induction.

(4) Add IPTG to the culture medium at a final concentration of 1 mmol/L, continuously culture in a shaker at 37 ℃ for 3–5 h.

(5) Harvest the bacteria from the culture medium by centrifugation at 10,000 g for 1 min using 1.5 mL Ep tube, discard the supernatant. Repeat the step once if necessary.

### 3.3.2  Identification of expressed target GST–fusion protein

(1) Identification of foreign protein expression: add 100 μL SDS–loading buffer in the bacteria pellet as control samples before and after induction, respectively, mix well, boil for 5 min, chill on ice for loading on SDS–PAGE(refer to Experiment 2, Chapter 8).

(2) After checking the expression level of foreign protein by SDS–PAGE, the specificity of the target protein could be identified by western blotting using an antibody against the target protein(refer to Experiment 3, Chapter 8).

### 3.3.3  Purification by affinity chromatography using Glu–agarose beads

(1) Add 0.5 mL lysis buffer to resuspend the bacteria pellet after induction, then add 50 μL lysozymes, mix well, shake for 1 h at 37 ℃ to disrupt cells. Remove the cell debris by centrifugation at

10,000 g for 1 min at 4 ℃. Transfer the supernatant into a new 1.5 mL Ep tube.

(2) Transfer 50% 50 μL suspending Glu–agarose beads into the supernatant in the Ep tube, mix well, gently shake for 5 min.

(3) Centrifuge, 3,000 g×1 min, to drop the beads at the bottom and discard the supernatant carefully.

(4) Add 1 mL wash buffer, mix well, centrifuge, 3,000 g×1 min, and carefully discard the supernatant. Repeat for 6–8 times.

(5) Centrifuge at 3,000 g for 2 min and sufficiently discard the supernatant.

(6) To elute the target protein, add 20 μL elution buffer containing 50 mmol/L reduced GSH into Glu–agarose beads. Mix and incubate at room temperature for 10 min, then centrifuge the mixture at 3,000 g for 2 min and carefully collect the supernatant in a new 1.5 mL sterilized tube.

### 3.3.4　Purity identification of target proteinby SDS–PAGE

Add an equal volume of 2×SDS–loading buffer into eluted solution, mix well, boil for 5 min, prepare to purity identification of the target protein by SDS–PAGE( refer to Experiment 2, Chapter 8). Loading samples include cells lysate before induction (10 μL), crude extract (10 μL) and purification products (15 μL), run the gel and analyze the results.

## 3.4　Results and discussion

(1) Results of SDS–PAGE ( Figure 21.2).

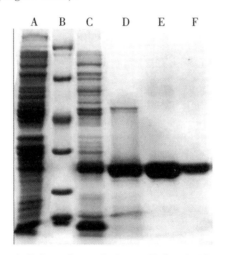

A. Before induction; B. Protein Marker; C. After induction and before purification; D. Low purity of protein after purification; E, F. High purity of protein after purification.

Figure 21.2　Results of SDS–PAGE

(2) Is there any other more accurate and sensitive method to identify the recombinant products?

(3) List the main merits and shortages of GST–tag, and thinking about other common tags like GFP–tag or His–tag?

# 3.5   Troubleshooting

The problems that may be encountered in induced expression and purification of the target gene and guidance for how to address major difficulties (Table 21.9).

Table 21.9   Troubleshooting in induced expression and purification of the target gene

| Problem | Potential causes | Solutions |
| --- | --- | --- |
| · No obvious target protein band after induction | · Failure of induction | · Check the · concentration of IPTG<br>· Prolong induction time<br>· Change expression system<br>· Expand culture volume |
| · Weak target protein band after induction | · Lower expression level | · Adjust the temperature and time of inducing |
| · Weak target protein band after glutathione Affinity purification | · The lower efficiency of purification | · Increase the duration of elution<br>· Increase the concentration of reduced glutathione in the elution buffer |

*Yu Hong, Du Fen, Shi Lei*

# Design of Experiment and Informational Experiment

# Chapter 22

# Design of Innovative Experiment

The design of innovative experiments is the design of any task that aims to test a hypothesis based on students' basic knowledge and experimental skills. The aim is to mobilize the students' initiative, enthusiasm, and creativity, and to cultivate the students' comprehensive ability to analyze and solve problems. An ordinary experiment is usually given, completed by the students themselves under certain experimental purposes and conditions; an innovative experiment focuses more on further stimulation of students' creative thinking and innovative consciousness, requiring the students to combine the knowledge, the interest of independent topics together.

**Description of innovative experiment**

(1) Fully self-designed innovative experiment

The experiment is designed by students themselves, including experimental subjects, materials, procedures and so on.

(2) Specific innovative experiment

The supervisor stipulates the scope of the experiment within limits.

(3) An improved and supplementary innovative experiment

Improve or supplement the original experimental scheme or add new creative experiments.

## Section 1   Selection of Experimental Subject

## 1.1   The principle of subject selection

### 1.1.1   The practicability of the subject

The subject should have practical values and the topic should be in accordance with the needs of social and scientific development.

### 1.1.2   The feasibility of the subject

The experiment must have certain subjective and objective conditions that can be completed. The project designer should fully analyze all the necessary objective conditions for the experiment, including the experimental tools, the source of the experimental objects, the duration of the experiment and also the funding.

### 1.1.3  The progressiveness of the subject

The project designer should demonstrate the progress and the problems that have been made of the subject. And the researchers can carry out the next scientific and innovative research on the basis of the previous results.

### 1.1.4  The scientificity of the subject

The subject should be based on scientific theories and certain facts. It can't contradict the basic scientific laws and theories that have been testified.

## 1.2  The procedure of subject selection

### 1.2.1  Raising questions

The most important thing in scientific research is to find out questions. During the basic research or clinical practice, there are always problems that are unclear or unsolved. When we find out and try to solve these problems, the tentative ideas of scientific research come out from our mind.

### 1.2.2  Literature review

Looking up the relevant references and monitoring the research progress related to the subject.

### 1.2.3  Making a hypothesis

A hypothesis, always tested by scientific methods, is based on previous observations that cannot satisfactorily be explained with the available scientific theories.

### 1.2.4  Experimental scheme of testing hypothesis

Designing a series of experiments to verify the feasibility of a hypothesis.

### 1.2.5  Naming the title of the subject

Based on the hypothesis and experimental scheme, the subject designer can determine the title of the subject. It should include three elements: the experimental object, the experimental factor, and the experimental effect.

## Section 2   Experimental Design

## 2.1  Elements of experimental design

### 2.1.1  Experimental objects

The objects of medical experiments usually include animals and human tissues.

### 2.1.2  Experimental factors

The experimental factor refers only to the independent variables, including the dose, concentration, and effect time of a certain component.

Interference factor: experimental animals' age, sex, and weight.

## 2.1.3　Experimental effects

The experimental effect is a term used to describe any of a number of subtle cues or signals from an experiment that affects the performance or response of subjects in the experiment.

## 2.2　The principle of experimental design

Considering the requirements of statistical analyses, the main principles of experimental design include randomization, proper controls, replicates, and equilibrium.

### 2.2.1　The principle of proper controls

Controls eliminate alternative explanations of experimental results, especially experimental errors, and biases. Controls are specific to the type of experiment being performed. Control groups include negative control, self−control, inter−group control, and positive control.

### 2.2.2　The principle of randomization

In the statistical theory of designing experiments, randomization involves randomly allocating the experimental units across the treatment groups.

### 2.2.3　The principle of replicates

One of the main principles of the scientific method means the ability to independently achieve non−identical conclusions that are at least similar when differences in sampling, research procedures and data analysis methods may exist. Both of reproducibility and replicability are among the main beliefs of the scientific method—with concrete expressions of the idea of such a method varying considerably across research disciplines and fields of study.

### 2.2.4　The principle of equilibrium

Equilibrium means that the interference factors that may affect the results of the experiment should be coincident. For example, animal experiments require the number, breeding, sex, age, weight, and color to be consistent, so that the experimental errors can be effectively reduced.

# Section 3　Experimental Implement

## 3.1　The determination of the experimental implement

(1) Select the appropriate technical procedure and give a clear technical flowchart.

(2) Record the specific experimental methods and matters that need attention for each experiment step in the lab notebook.

## 3.2 Preparation of experimental materials

The experimental materials include samples, reagents, instruments, and equipment. It is specific to see the contents of the relevant chapters of this textbook.

## 3.3 Experimental record

(1) The phenomena observed in the experiment.
(2) The record of the original experimental data.

# Section 4　Experimental Summary

The discussion should be based on the experimental results, theoretically analyzing and comparing. The basic contents of the discussion include as follows.

(1) The experimental results are discussed with the existing theories. The relationship between general rules and particularity rules in the experiment is emphasized.

(2) What new problems are prompted by the results of the experiment?

(3) Put forward some points for attention and improvement in future experiments.

*Sun Wei, Yang Guang*

# Chapter 23

## The Basic Ideas of Designing Research Experiments

This chapter mainly introduces the basic methods and key points for designing research experiments to assess nucleic acids, proteins, enzymes, and other molecules.

## Section 1    Experiment Design of Nucleic Acid Research

The basic process of nucleic acid research includes the extraction, separation, and purification of nucleic acid, followed by studying its properties, structure, and function.

## 1.1    Basic route of nucleic acid research

The essence of nucleic acid research is to use gene manipulation technology to conduct a series of studies on the structure and function of a gene. Generally, gene manipulation technology refers to all kinds of technology involving DNA or RNA manipulation, the core of which is genetic engineering and other relative technologies. This method mainly includes "cutting" (cutting DNA with restriction enzymes), "linking" (using DNA ligases to linktwo DNA fragments), "transfer" (gene transfer techniques, e. g., DNA transformation and transfection of recipient cells), "amplification" (amplification of target DNA in a host/vector system or *in vitro* PCR amplification), and "hybridization" (southern blotting of DNA, northern blotting of RNA, etc. ).

The technical process of gene manipulation includes: ① gene isolation (genomic DNA/total RNA extraction, purification, and amplification of the target gene); ② gene copy number analysis; ③ gene structure (sequence) analysis and modification; ④ gene cloning; ⑤ gene expression and identification; ⑥ gene functional analysis(Figure 23. 1).

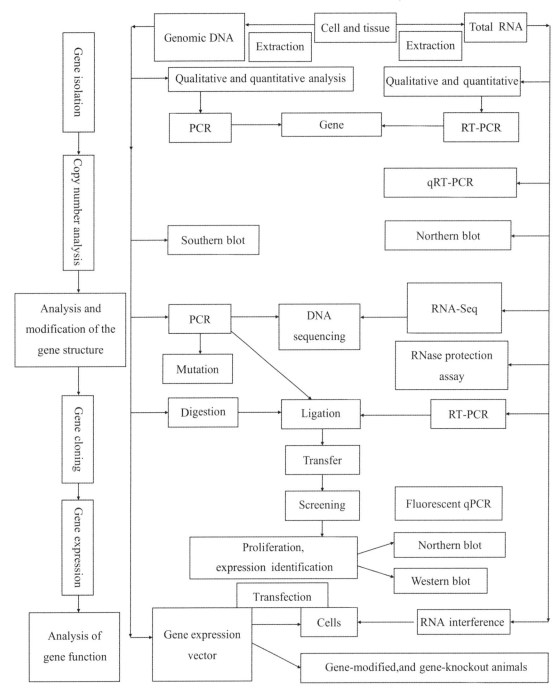

Figure 23.1 Flow chart of gene manipulation technology

# 1.2 Key points of experiment design for nucleic acid research

## 1.2.1 Isolation and purification of nucleic acids

Most of the eukaryotic DNA (95%) is present in the nucleus, whereas the rest (5%) is found in organelles such as the mitochondria and chloroplasts. However, RNA mainly exists in the cytoplasm ( about 75% ), with 10% of it distributed in the nucleus and 15% in organelles.

(1) Selection of material

Before the isolation and purification of nucleic acids, the first aspect to consider is the selection of appropriate material of nucleic acid extraction, which is conducive to the experiment's success. For example, some of the nucleic acid content is very low, unstable, easily destroyed, or denatured; and the same material at different growth stages contains different nucleic acid content, such as mRNA expression.

The basic principles of material culture and selection include: ①having a plentiful source of low−cost materials; ②obtaining a high concentration of the target nucleic acid with good stability, simple structure, and easy separability, extraction, and purification and with an accessible process. If specific circumstances occur during the experiment, the main problem could be understood to reach better choices after comprehensive consideration.

(2) The process of isolation and purification

1) Cell disruption: the general principle of cell disruption is that it should not influence the structure or function of the nucleic acid; that is, denaturation or inactivation of the target nucleic acid should be avoided as much as possible.

Cell disruption can be performed using mechanical, physical, chemical, or enzymatic methods. Mechanical methods include grinding and homogenization. Physical methods utilize various physical factors, such as temperature, pressure, and osmotic pressure, as well as ultrasonic crushing, including the squeezing method, ultrasonic method, and freezing and thawing method, as well as the low−permeability pyrolysis method, among other methods. Chemical methods utilize chemical reagents to destroy the cell membrane structure with organic solvents and surfactants. Enzymatic hydrolysis methods are used to destroy the outer structure of cell by adding enzymes.

Table 23.1 shows the applications and characteristics of several cell disruption methods.

Table 23.1   Comparison of the main methods of cell disruption

| Method | Application | Feature |
| --- | --- | --- |
| Grinding | Bacteria, yeast, and some plants and animal tissues | Mild conditions, suitable for laboratories |
| Homogenization | Bacteria, yeast, and some plants and animal tissues | Severe conditions, requiring low temperatures |
| Crushing | Bacteria, yeast, and plant tissues | Mild conditions, complete cell disruption |
| Ultrasonic | Bacteria, yeast | Easy operability, suitable for laboratories |
| Freeze/thaw | Cell culture | Suitable for the extraction of very stable proteins |
| Hypotonic lysis | Red blood cells, bacteria | Easy operability, poor versatility |
| Organic solvent | Bacteria, yeast | Pay attention to the stability of the target protein |
| Surfactant | Tissues, cells culture | Pay attention to the stability of the target protein |
| Enzymolysis | Bacteria, yeast | Selectivity, poor versatility |

2) Extraction: in this process, an extraction solution is added to extract the nucleic acid under mild conditions. Nucleic acids are generally soluble in water and dilute acidicor alkaline solutions, but insoluble in organic solvents, such as ethanol, acetone, and butanol. Therefore, different solvents can be used to separate nucleic acids.

3) Isolation and purification of DNA/RNA: DNA and RNA are purified using protein denaturants ( e. g. , phenol, chloroform) , detergents ( e. g. , SDS) , and protease. After the protein is removed by centrifu-

gation, the nucleic acid is separated from the protein and remains in the aqueous phase ( supernatant). Then, absolute ethanol is added to the aqueous phase to precipitate the nucleic acid. Finally, the nucleic acid is purified using chromatography or ultracentrifugation. Notably, RNase A is used for degrading RNA during DNA extraction. In addition, while extracting RNA, the RNA samples are treated with DNase Ⅰ to remove the genomic DNA contamination.

4) Identification: after the extraction of genomic DNA or total RNA, qualitative, quantitative, and purity analyses of DNA/RNA by electrophoresis and determination of UV absorbance are usually required.

## 1.2.2　Gene isolation

The following are examples of methods used for obtaining a gene of interest from purified nucleic acids: ① genomic DNA cutting with restriction endonucleases and separation by electrophoresis or ultracentrifugation; ② PCR amplification of genomic DNA; ③ RT-PCR amplification using total RNA as a template.

## 1.2.3　Analysis of the gene copy number

Southern blotting, which is based on the location and frequency of the probe signal to determine the gene copy number, or real-time fluorescence quantitative PCR is normally utilized to detect the gene type and analyze the copy number.

## 1.2.4　Analysis and modification of gene structure

The primary structure of a gene ( i. e., base sequence) can be directly analyzed using DNA sequencing. In addition, the primary structure of mRNA can also be determined using RNA-Seq ( sequencing of cDNA after reverse transcription) or RNase protection assay ( RPA).

Genetic modification refers to the alteration of the primary structure of a gene by changing its base sequence. The most common technique used for genetic modification is site-directed mutagenesis, which involves changing specific bases in a gene sequence. Currently, PCR is considered a popular technique for carrying out gene mutations.

## 1.2.5　Gene cloning

Gene cloning, also known as molecular cloning or DNA cloning, is accomplished using recombinant DNA technology. The basic steps of this process are as follows: separation, digestion, ligation, transfer, and screening. Chapter 21 discusses the specific experimental methods and principles associated with this technique.

## 1.2.6　Gene expression and identification

To obtain the expression products ( protein), the gene of interest is first cloned into an expression vector and then introduced into a corresponding expression system, for example, either a prokaryotic expression system ( the most common being the *E. coli* expression system) or a eukaryotic expression system ( the yeast expression system, the mammalian cell expression system).

Gene expression can be identified mainly at the RNA level using semiquantitative RT-PCR, fluorescence qPCR, and northern blotting and at the protein level by western blotting.

## 1.2.7　Analysis of gene function

Gene function can be inferred by the function or activity of the gene expression products. However, transferring the gene back into the organism for research should be more accurate.

( 1) Molecular level

After verifying the levels of DNA and RNA, it is necessary to study the level of protein and to perform

DNA co-analysis in gene functional studies. Chromatin immun oprecipitation sequencing (ChIP-seq) is a technology that has been developed by combining second-generation sequencing and proteomics. Generally, this technique is specifically used to enrich DNA fragments bound by the target protein, followed by the purification of DNA fragments and using them to construct a library for high-through put sequencing. Finally, by accurately locating millions of sequence tags on the genome, genome-wide DNA information related to histones and transcription factors can be obtained.

(2) Cellular level

At the cellular level, gene function is directly studied by transfecting genes into cells. The key aspect here is to construct a gene transfection cell model. Notably, the requirement for gene-transfected cells differs depending on the purpose of the experiment. If only transient expression of the target gene is required, then there will be no need to screen the transfected cells. After transient transfection, the expression of the gene of interest reaches its peak at 48-72 h. However, because of the limited number of gene-transfected cells (without extensive amplification), transfected cells soon become nondominant and eventually disappear. Stable transfection is necessary to obtain cell lines with sustained target gene expression. For example, when transfected cells are cultured under pressure selection (e. g., drug resistance screening) through the lentivirus expression system, cells without target genes die during the selection process, and hence cells with genes survive and cell lines with stable target gene expression are obtained. To avoid gene toxicity of cells after transfection, it is necessary to regulate the expression of the target gene. An expression vector containing an inducible promoter should be chosen for transfection. However, gene expression keeps shutting down in transfected cells until an inducer is added, hence representing an inducible type of expression.

(3) Animal model level

When performing gene function studies in an animal model, the gene of interest is first usually cloned into a suitable vector. Then, the gene is transferred into the chromosomes of the recipient cells using specific techniques, such as transgenic technology or gene targeting. Following this, research can be performed on transgenic animal models or knockout animal models. In recent years, CRISPR/Cas9 technology has been utilized to edit genomic DNA using guide RNA (gRNA) and Cas9 protein.

# 1.3    Case study and practice of nucleic acid experiment design

Title: nucleic acid extraction, identification, quantification, and alkaline phosphatase gene cloning and expression (*note: nucleic acids including genomic DNA and RNA*).

In eukaryotic cells, nucleic acids (genomic DNA and RNA) and proteins are usually present in the form of nuclear protein and DNA is mainly present in the nucleus. DNA is soluble in water and solutions with high salt concentrations(e. g., 1 mol/L NaCl) but poorly soluble in solutions with low salt concentrations(e. g., 0. 14 mol/L NaCl). When organic solvents are used in extraction, DNA can be dissolved in the aqueous phase and ethanol can be used to precipitate it, whereas RNA can be dissolved in an alkaline solution, to adjust its pH to neutral, and ethanol can be added to precipitate it. Tissue cells contain small fat-soluble and acid-soluble molecules. Moreover, RNases are widely present in cells.

# Section 2    Experiment Design of Protein Research

Generally, different cells contain thousands of different proteins, often present in complex mixtures. The basic process of protein research is as follows: protein extraction, crude separation, and then purification, followed by qualitative, structural, and functional analyses.

# 2.1 Basic framework of protein research

Protein research focuses on either a single protein or whole proteins (proteome) as the research object. The basic line of protein research is shown in Figure 23.2.

## 2.1.1 Single-protein research

In this method, the target protein is first directly separated from the protein extraction solution to obtain a high-purity sample. Alternatively, various protein isolation and purification methods can be used to sequentially remove other protein impurities from the protein extraction solution. This is then followed by studying the physical and chemical properties of the target protein, its structure and biological function, and the relationship between its structure and function.

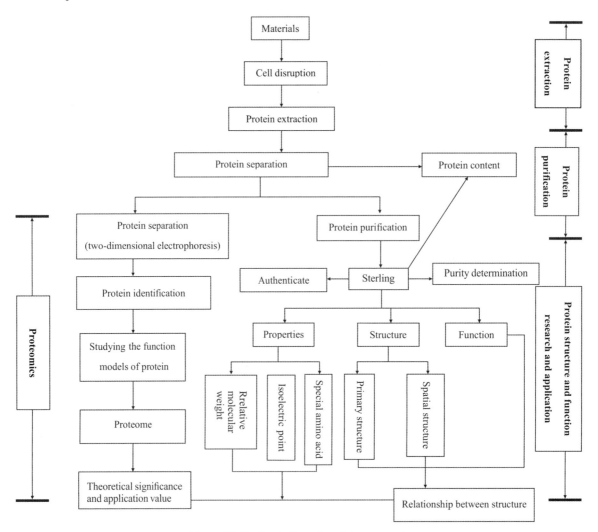

Figure 23.2 **Basic line of protein research**

## 2.1.2 Proteomics research

Generally, it is possible to analyze the protein expression profile under various conditions and the function of relative important proteins using high-resolution protein separation techniques ( e. g. , two-dimensional electrophoresis) and protein identification techniques ( e. g. , high-performance mass spectrometry) to establish a proteome database. Moreover, the biological significance and mechanism of proteins in biological organisms can be studied under at whole-proteome level.

# 2.2   Experimental methodology of protein research

## 2.2.1   Protein extraction

The key problem in protein extraction, separation, and purification is to maintain the activity of the target protein.

(1) Cell disruption

Intracellular proteins are either freely located in the cytoplasm, tightly bound to organelles, or distributed in the nucleus. Therefore, cells must first be broken up to release these proteins before extraction. The general idea when selecting a cell disruption method is to avoid the denaturation or inactivation of the target protein.

(2) Protein extraction

After cell disruption, proteins are extracted using an extraction buffer under mild conditions. Most proteins are soluble in water, dilute salt solutions, and dilute acidic or alkaline solutions; a small number of lipid-bound proteins are soluble in ethanol, acetone, butanol, and other organic solvents. Therefore, each protein is extracted using a different solvent depending on its type.

During protein extraction, the following precautions need to be taken: ①the extraction temperature generally needs to be low ( about 4 ℃) ; ②the pH of the extraction buffer should be within the stable range of the isoelectric point of the target protein; ③a neutral salt solution with low ionic strength should be added to the extraction buffer to maintain protein activity; ④the extraction buffer should contain reducing agents ( e. g. , β-mercaptoethanol) , metal-ion-chelating agents ( e. g. , EDTA) , protease inhibitors, and other ingredients to maintain the stability of the protein of interest; ⑤the amount of extraction buffer should be one to five times the volume of the raw materials and the extraction buffer needs to be mixed well to facilitate protein dissolution.

## 2.2.2   Determination of the protein content

Protein content can be determined using methods dependent on the protein's physical properties ( UV absorption, refractive index, specific gravity, etc. ) as well as chemical methods ( e. g. , nitrogen determination, biuret reaction, and Folin's phenol reaction) . According to the principles of chemical methods, determination of protein content is classified into two categories: ①initial determination of the nitrogen content followed by calculating the sample solution's protein content, ②the addition of achemical reagent to the protein solution to perform a color reaction and ultimately using a colorimetric method ( spectrophotometry) to determine the absorbance. For a comparison of the advantages and disadvantages of the protein content determination and specific experimental principles and procedures, please refer to Chapter 7.

## 2.2.3   Purification of proteins

Generally, the principles and operation of protein separation and purification methods depend on the protein's physical and chemical properties. Detail methods are referred to in relevant professional books and papers. The following list briefly introduces the basis upon which each protein separation and purification method is selected.

(1) Based on the protein's molecular size: dialysis, ultrafiltration, and gel filtration chromatography.

(2) Based on the protein's solubility difference: salting out, organic solvent precipitation, isoelectric point precipitation, and selective precipitation (based on the protein's sensitivity to temperature, heavy−metal salts and alkaloids for selective precipitation, or the use of specific binding between the protein and protein−ligand for purifying the target protein).

(3) Based on the protein's charge difference: PAGE, capillary electrophoresis, isoelectric focusing, ion−exchange chromatography, and chromatography focusing.

(4) Based on the protein's selective adsorption strength: hydroxyapatite column chromatography and hydrophobic interaction chromatography

(5) Other methods: affinity chromatography, high−performance liquid chromatography (HPLC), and microcrystallization.

## 2.2.4  Purity identification of proteins

Currently, isoelectric focusing, capillary electrophoresis, SDS−PAGE, gel filtration chromatography, HPLC, mass spectrometry, and crystallization, among other methods, are used to identify the purity of proteins. When only one band or peak is present in the results, the protein sample is considered to be pure.

## 2.2.5  Research on the chemical properties of proteins

(1) Relative molecular weight determination of proteins: SDS−PAGE is described in Experiment 2, Chapter 8, in detail. In addition, ultracentrifugation, gel filtration, osmometry, chemical composition determination, capillary electrophoresis, and mass spectrometry are also used to determine the relative molecular weight of proteins.

(2) Isoelectric point determination of proteins: isoelectric focusing (PAGE isoelectric focusing, capillary isoelectric focusing electrophoresis, etc.) is used to determine the isoelectric point of proteins.

(3) Identification of specific amino acids in proteins: some amino acids in protein molecules exhibit special color reactions; these special amino acids can be easily identified by relative reactions. For example, cysteine reacts with nitroprusside, yielding a red color.

(4) Identification of the target protein: SDS−PAGE, western blotting, enzyme−linked immunosorbent assay (ELISA), co−immunoprecipitation, HPLC, and mass spectrometry are used to identify target proteins.

## 2.2.6  Protein structure analysis

(1) Determination of a protein's primary structure: chemical analyses (i.e., sequence analysis of the primary structure of aprotein) and deduction (by determining the nucleotide sequence encoding the protein to infer the amino acid sequence of the corresponding protein) are used to determine the primary structure of proteins.

(2) Determination of a protein's spatial structure: circular dichroism (CD), X−ray diffraction, nuclear magnetic resonance (NMR) spectroscopy, fluorescence spectroscopy, and protein structure prediction (bioinformatics−based methods, predicting the three−dimensional structure from the amino acid sequence of a protein) are among the methods used to determine the spatial structure of proteins.

## 2.2.7  Protein function research

(1) Chemical modification of proteins: this method mainly includes structural changes inside−chain groups and the main chains of proteins. Structural changes lead to changes in the biological activity of proteins. Therefore, protein chemical modification is an important method to study the relationship between the structure and function of a protein. Chemical modification of protein side−chain groups can be mainly achieved via the chemical reaction of selective reagents or affinity labeling reagents with specific functional groups on them.

(2) Interactions between proteins: the main research methods used to determine the interactions be-

tween proteins include the yeast two-hybrid method, phage display technology, chemical cross-linking, and tag-fusion protein system screening.

(3) Protein engineering: in this method, proteins are modified and redesigned to produce new proteins. The basic steps of this method are to design the new expected structure, identify the corresponding amino acid sequence, translate it into a nucleotide sequence, and then synthesize a new protein by biosynthesis (from gene to protein) or artificial synthesis. This process is called reverse biology.

(4) Proteomics: proteomics is the study of the expression patterns of all proteins as well as the functional patterns and interactions between proteins under specific conditions at the overall-protein level.

# 2.3　Selection of experimental methods of protein research

Generally, choosing an appropriate method for protein research depends on the aim of the experiment, the characteristics of the experimental materials, the target protein's physical and chemical properties, and operation simplicity, among other factors.

For proteins with unclear properties, it is generally required to select two or more different experimental methods from different principles and levels for comparative research and mutual verification. The principle, scope of application, and main features of each method should be fully considered to make reasonable arrangements.

The first choice of a protein separation and purification method should be extensive, rapid, and helpful in reducing the sample volume and simplifying the subsequent processes (e. g. , precipitation method, ultrafiltration). On the other hand, the remaining part of the experiment requires accurate, time-consuming, and sampleless methods (chromatography, electrophoresis, etc.).

Gel filtration and SDS-PAGE are generally selected to determine the relative molecular weight of proteins. Gel filtration is used to measure the relative molecular weight of the native intact protein, and SDS-PAGE is used to determine the relative molecular weight of different subunits of the denatured protein. Using these two methods to determine one protein at a time can help roughly determine the subunit type and relative molecular weight of a protein.

Moreover, operable experimental methods using only necessary equipment should be chosen. For example, to determine the protein content, compared to the Kay nitrogen method and amino acid analysis, methods employing Coomassie Brilliant Blue and Folin's phenol reagent should be the first choice. In terms of the precipitation method, salting out combined with isoelectric point precipitation is usually selected.

# 2.4　Case and practice of protein experiment design

## 2.4.1　Analysis of a case study and practical experiment design

(1) Title: Extraction, Isolation, and Purification of Casein from Milk and Properties Research.

(2) Technical route of experiment design (Figure 23.3).

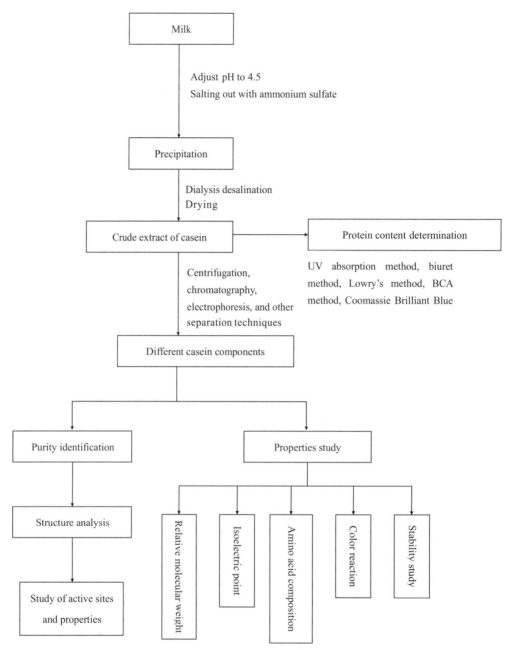

Figure 23.3  Technical route of casein experiment

## 2.4.2  Practice of protein experiment design

(1) Title: Extraction, Isolation, Purification, and Quantitative Determination of Protein from Rat Brain Tissue.

(2) Known: the molecular weight of this protein is about 17 kDa, and its isoelectric point is in the range 3.9–4.1. This protein is resistant to high temperatures and retains 80% of its activity even after being heated at 90 ℃ for 3–4 min. In addition, this protein contains hydrophobic residues, which become exposed after the protein binds with $Ca^{2+}$.

## Section 3   Experiment Design of Enzyme Research

### 3.1   Basic route of enzyme research

Enzymes are proteins that efficiently catalyze specific substrates and are considered the most essential catalyst to catalyze a variety of metabolic reactions in an organism. Enzyme research is comprised of three branches: enzyme theory study, enzyme engineering, and enzyme application research. The basic route of enzyme research is shown in Figure 23.4.

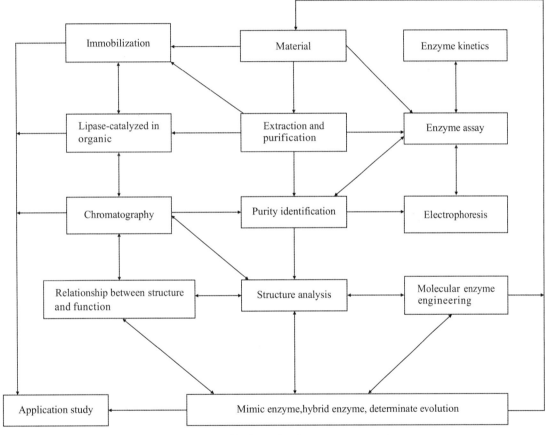

Figure 23.4   Basic route of enzyme research

### 3.2   Key points of experiment design for enzyme research

The basic process of enzyme research is slightly different from that of protein research and is mainly comprised of the following steps: enzyme protein extraction, isolation, and purification; purity identification; enzyme activity assay; and enzyme kinetic analysis.

## 3.2.1 Materials and processing

(1) Sources of enzymes

Enzymes are mainly produced from microbial fermentation as well as plant and animal cultures, with microbial material being the most important source of enzymes. Owing to the wide variety of microorganisms and their short life cycles and ease of cultivation, commercial enzymes are mostly produced from microorganisms. Great progress has also been reached in the utilization of plant cell culture technology for enzyme production, in contrast to the use of animal cell culture technology to produce enzymes (collagenase, urokinase, etc.), which is expensive and requires stringent culture conditions.

With the exception of enzymes that are microbial extracellular enzymes or enzymes located in animal and plant body fluids, most enzymes exist in cells and are localized in subcellular organelles, the cytoplasm, mitochondria, lysosomes, microsomes, and peroxisomes. For example, succinate dehydrogenase is distributed in the mitochondria, where as catalase is distributed in the peroxisomes. Therefore, when studying enzymes, it is important to consider the regionalization of enzyme distribution. Furthermore, when mitochondrial enzymes are regarded as research targets, mitochondria should be isolated as enzyme material.

(2) Raw material processing

Two processing methods can be used to extract enzymes from microorganisms: ①extracellular enzymes can be secreted into a culture medium by microbial cells, and the culture medium of the microbial cells can then be collected and centrifuged to remove the precipitate bacterial component; ②most enzymes are intracellular, and therefore it is necessary to break their cell membrane or cell wall. However, the cytoderm of yeast is hard to break, which increases the cost and difficulty of enzyme extraction. For details on cell disruption, please see the section on protein experiment design for nucleic acid reseath.

When using plant material, it is necessary to break the cytoderm, degrease the material, and pay attention to the plant species, as well as changes in growth and development, along with seasonal changes. On the other hand, when using animal tissue, organ tissues rich in the active ingredient must be chosen as the raw material. Mincing should be completed first, followed by degreasing and other processes. Notably, some enzymes in animals exist as zymogens. For example, trypsin is present in the pancreas of animals as trypsin zymogen. Therefore, corresponding measures should be taken to activate zymogen after extraction.

When the research object is a living organism, the material should be selected according to the purpose of the experiment. For example, when exploring a series of life phenomena associated with one or several enzymes, such as biological development, growth, and disease, the most important aspect of experiment design is to consider drawing material from which life stage, which suborganelles, and what growth environment. For example, the blood sample drawn from a patient to determine the activity of alanine aminotransferase as an important indicator of liver function can only be drawn from the patient in a fasting state. Pretreated materials should be frozen if not tested immediately, and biomacromolecules that degrade easily should be prepared with fresh materials.

## 3.2.2 Enzyme extraction

The purpose of enzyme extraction is to separate enzymes from cells or other enzyme-containing raw materials. In this process, enzyme-containing raw materials are processed with an appropriate solvent or solution to fully dissolve enzymes. One of the key issues in this process is to maintain enzyme activity.

(1) Solvent or solution selection depends on the enzyme's dissolution characteristics.

1) When extracting enzymes that are easily soluble in water or dilute salt solutions, dilute salt solutions and buffers are commonly used because enzymes in solution are stable and soluble. Organic solvents can be used when extracting enzymes that are nonpolar orsoluble in these types of solvents.

2) Dilute salt solutions, such as 0.15 mol/L NaCl, and other neutral salt solutions can promote enzyme dissolution (salting-in effect). Salt ions bind to proteins to protect them from denaturation. During enzyme

extraction, the buffer solution used is usually a $0.02-0.05$ mol/L phosphate and carbonate isotonic saline solution. When a dilute acid or alkali is used in extraction, it is necessary to prevent irreversible changes in the protein conformation that might occur as a result of excessive acid or alkali. Generally, alkaline proteins are extracted using acidic extract solutions, where as acidic proteins are extracted using alkaline extract solutions.

3) Some proteins and enzymes that can bind to lipids are insoluble in water, dilute salt solutions, dilute acids, or dilute alkalis. Some organic solvents, such as ethanol, acetone, and butanol, which have a certain degree of hydrophilicity and strong lipophilicity, can be used for extracting these enzymes, a process that requiresa low temperature. Notably, butanol extraction is suitable for the extraction of some proteins and enzymes that are closely associated with lipids. This method is applicable under a wide range of temperatures and pH values and is suitable for animal, plant, and microbial materials.

(2) Buffers provide a suitable environment in enzyme extraction to maintain enzyme activity. When determining enzyme activity, an activator should also be added to provide optimal enzyme reaction conditions.

(3) As previously described, the solvent or solution used should be $1-5$ times the volume of the raw material. To facilitate enzyme dissolution, it is important to mix the solution well during extraction. However, if the volume of the solution is too large, the enzyme can be diluted, which is not conducive to subsequent purification steps, and the enzyme is easily deactivated.

(4) The temperature used should be determined according to the nature of the active ingredient during extraction. Generally, high temperatures cause the dissolution of enzymes and shorten the extraction time but result in protein denaturation. Therefore, enzyme extraction is generally performed at low temperatures (below 5 ℃).

## 3.2.3　Isolation, purification, and purity identification of enzymes

Enzymatic isolation and purification is a step-by-step separation process between the target enzyme and hybrid protein in the enzyme crude extract. This process is performed using specific methods.

Enzymatic activity, protein concentration, and specific activity should be determined after isolating and purifying an enzyme. The efficiency of the purification method, for example, extraction efficiency and purification fold should be concerned with.

Electrophoresis, ultracentrifugation, and HPLC, among other methods, are often used for purity identification of enzymes, in addition to PAGE, isoelectric focusing, and capillary electrophoresis, which are also widely utilized. The presence of only one band or a single symmetrical peak is indicative of a pure enzyme.

## 3.2.4　Determination of enzyme activity and enzyme kinetic analysis

Determining the enzyme activity is the most important aspect in the whole experimental process. In this process, accuracy, convenience, and speed are required. Generally, three main types of assays exist: ①direct assay (direct detection of changes in the substrate or product content), ②indirect assay (indirect detection of changes in the substrate or product via non-enzyme-assisted reactions), and ③conjugation assay (which does not directly detect the substrate or product of the original reaction but indirectly reflects the changes in the substrate or product of the enzyme to be assayed by coupling to another enzyme or accessory enzyme for the detection of the enzymatic reaction products).

Enzyme kinetic analysis generally includes the following aspects: ①effect of substrate concentration on the enzyme activity; ②effect of pH on enzyme activity; ③effect of temperature on enzyme activity; ④effect of metal ions on enzyme activity; ⑤effect of NaCl on enzyme activity; ⑥effect of EDTA on enzyme activity; ⑦effects of other factors, such as the impact of thiol modifiers on enzyme activity.

## 3.3   Case and practice of enzyme experiment design

### 3.3.1   Case study of experiment design

(1) Title: Extraction, Activity Determination, and Enzyme Kinetic Analysis of Amylase.

(2) The route of experiment design ( Figure 23. 5)

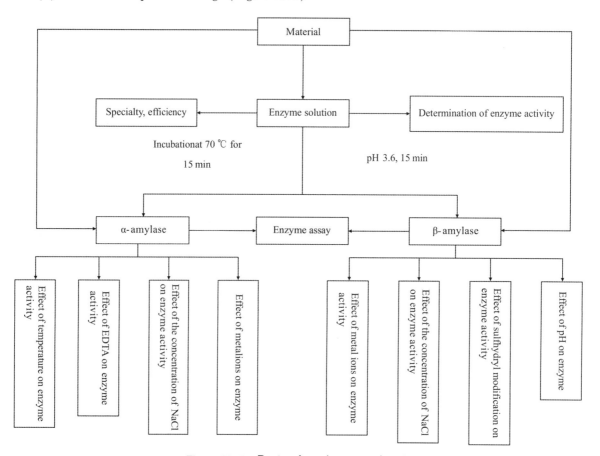

Figure 23.5   **Route of amylase experiment**

### 3.3.2   Practice of experiment design

(1) Title: Purification, Crystallization, and Enzyme Kinetic Analysis of Porcine Trypsin.

(2) Extraction, enzyme activity determination, crystallization, and relative molecular weight determination of egg white lysozyme.

# Section 4   Experiment Design of Carbohydrates, Lipids, and Vitamins

## 4.1   Experiment design of carbohydrates

Sugars, also known as carbohydrates, are chemically classified as polyhydroxy aldehydes or ketones, and their polymers and derivatives. The main forms of carbohydrates found in the human body include monosaccharides, oligosaccharides, polysaccharides, and glycoconjugates. According to the number of carbon atoms, monosaccharides are divided into hexoses, pentoses, trioses, and so on. On the other hand, according to the group, monosaccharides are divided into aldoses (e. g. , glucose) or ketoses (e. g. , fructose). Disaccharides, such as maltose, sucrose, and lactose, are considered the most common oligosaccharides. Polysaccharides mainly include glycogen, starch, and cellulose. Glycoconjugates include glycoproteins, proteoglycans, and glycolipids.

### 4.1.1   Basic processes and key points of carbohydrate experiment design

The basic process of carbohydrate experiments is comprised of sugar extraction, qualitative and quantitative determination, and structure analysis, functional research, and application.

(1) Carbohydrate extraction and purification

With the exception of cellulose and some glycoconjugates, most polysaccharides are water–soluble. Carbohydrates can be extracted by boiling with hot water followed by precipitation using 95% ethanol (the extraction conditions can be optimized via an orthogonal test of the extraction temperature, time, and solid–liquid ratio, among other factors). Protease, Sevag method, and organic solvent denaturation are used to remove impurities such as proteins in the crude extract. For example, in Experiment 1 Chapter 15 of liver glycogen extraction, trichloroacetic acid is added at a low concentration to denature the protein, destroy the enzymes in the liver, precipitate the protein, and separate the glycogen from other components, such as proteins.

According to the study purpose, crude extracts should be purified using column chromatography (e. g. , DEAE–Sephadex A–25 and Sephadex G–200). Extraction and purification of glycoconjugates, such as glycoproteins and glycolipids, should follow the principles and methods of protein and lipid extraction. After the purification of a glycoprotein sample, glycan is released in two ways: chemical release and enzyme release. Chemical release mainly comprises hydrazinolysis and $\beta_2$ elimination. Glycan needs to be hydrolyzed into monosaccharides for further qualitative and quantitative analysis of the components.

(2) Qualitative and quantitative determination of carbohydrates

Carbohydrates are hydrolyzed into monosaccharides using acids, followed by qualitative and quantitative determination, taking advantage of the reducibility of monosaccharides. The purpose of qualitative analysis is to determine whether the assessed sugars are pentoses, hexoses, aldoses, or ketoses. Generally, the reducibility of monosaccharides is influenced by the number of carbon atoms, the number of hydroxyl groups, and the difference in the carbon atoms' configuration due to the orientation of the hydroxyl groups. After treatment using a nitroreagent, copper reagent, sulfuric acid, or hydrochloric acid, colorimetry can be used for qualitative and quantitative determination. More details can be found in the relevant specialized literature. For example, in Experiment 1 Chapter 15, the principle of spectrophotometry is used to assess glycogen using the anthrone–sulfuric acid method.

(3) Carbohydrate structure analysis

Recently, following genomics and proteomics, glycomics has become a new frontier and hot spot in life science research. Glycomics is a science that analyzes the overall composition and activity rules of carbohydrates by studying carbohydrates expressed in specific time and space or under specific physiological and pathological conditions. Research on the glycan structure, function, and regulation is considered one of the important parts of glycomics. Glycan has different lengths, participates in various life processes, and has different forms in glycoproteins and lipoproteins.

Structural analys is of glycan chainsis divided into two parts: primary−structure determination and secondary−structure analysis. The following part is a description of the primary−structure determination of glycan. Generally, the methods used for primary−structure determination commonly include: ①physical methods( infrared spectroscopy, NMR) ; ②chemical methods( methylation, hydrolysis, and hydrazinolysis) ; ③ biological methods( glycosidase hydrolysis, immunological method) . Table 23. 2 shows more specific methods of selection.

Table 23.2　Contents and common methods for the determination of the primary structure of glycan chains

| Content | Methods |
|---|---|
| Relative molecular weight | Gel filtration, mass spectrometry, vapor pressure, etc. |
| Composition and proportion of monosaccharide | Partial acid hydrolysis, complete hydrolysis, paper chromatography, thin−layer chromatography, gas chromatography |
| d− or l−configuration analysis | Infrared spectroscopy |
| Monosaccharide residue sequence | Selective acid hydrolysis, sequential hydrolysis of glycosidase, NMR |
| α− or β−anomeric isomerization of glycosidic bonds | Infrared spectroscopy, glycosidase hydrolysis, NMR |
| Replacement of hydroxy group | Methylation−gas chromatography, periodic acid oxidation, mass spectrometry, NMR |
| Connection of sugar chains and nonsugar moieties | Monosaccharide and amino acid composition, dilute acid hydrolysis, hydrazinolysis |

( 4) Functional research and application

Among carbohydrates, polysaccharides play a vital role in immunomodulation and free radical scavenging and have antiviral, antitumor, and antioxidative effects. Therefore, experiments should be designed according to the corresponding content using cell or animal models. In addition, functional research experiments on glycoproteins and other carbohydrate complexes should be designed according to the corresponding types of substances, such as proteins.

## 4.1.2　Case study and practice of carbohydrate experiment design

( 1) Title: Extraction of Polysaccharides and Their Antioxidative Effect.

( 2) The materials chosen include aloe, *Rehmannia*, bitter melon, and *Polygonatum*.

( 3) Requirement: qualitative and quantitative determination.

# 4.2   Experiment design of lipids and vitamins

## 4.2.1   Experiment design of lipids

(1) Extraction of lipids

Lipids are not water-soluble; therefore, their extraction methods should involve organic solvents, solubility method, saponification, mechanical pressing, and other special techniques. ①Organic solvent extraction is a common method of lipid extraction in which nonpolar solvents (chloroform, benzene, ether, etc.) are often used to separate nonpolar hydrophobic lipids. During experiments, alcohol, as a combination agent, is commonly added to the organic solvent, for instance, a chloroform/methanol (2 : 1, v/v) mixture. Notably, the water endogenous to tissue constitutes the third component of this extraction reagent: chloroform/methanol/water (1 : 2 : 0.8, v/v/v). ② The solubility method depends on the different solubilities of extracted lipids in different solvents. ③Saponification depends on the hydrolysis of triglycerides into water-soluble sodium or potassium fatty acids (i. e., soap, glycerin). After obtaining the saponification solution, fatty acids are separated using acid treatment. ④Mechanical crushing uses physical pressure to squeeze oil out of the broken cells.

(2) Isolation and analysis of lipids

1) Chromatography: silica gel column chromatography, ion-exchange chromatography, thin-layer chromatography, and so on.

2) HPLC, gas chromatography.

(3) Determination of lipid structure

Determining the structure of a lipid includes determining the length of the hydrocarbon chain and the position of the double bond. After the lipid reacts with the appropriate reagents, mass spectrometry can be used for detection.

(4) Experiment design

Title: Preparation and Refining of Soybean Phospholipids.

## 4.2.2   Experiment design of vitamins

The basic process of vitamin experiments includes extraction and purification, qualitative and quantitative determination, and functional studies of vitamins. Generally, the analysis of vitamins in samples is divided into the following steps: ① Release of vitamins(acid/alkali treatment or enzymatic decomposition). ②Extraction of vitamins(solvent extraction). ③Purification(excluding interfering substances because vitamins are often combined with other substances; thin-layer chromatography, column chromatography, and HPLC can be used). ④Qualitative and quantitative determination[ the appropriate method should be selected by considering the sample conditions (generally low content) and the characteristics of the method itself]. Finally, functional studies, such as antioxidation experiments of vitamins, can be performed by comparing vitamins with other antioxidants.

*Li Ling, Lin Guanchuan, Yu Hailang*

# Chapter 24

## The Basic Method of Bioinformatics

## I. The concept of bioinformatics

Bioinformatics is the combination of computational science and biology (life science). It focuses on the studies of the collection, processing, storage, transmission, analysis and interpretation of biological information. It reveals the mysteries contained in numerous and complex biological data by comprehensively utilizing biology, computational science and information technology.

## II. The research content of bioinformatics

There are many research objects in bioinformatics. In general, anything with biological significance is the subject of this discipline. The research objects can be divided into three types: nucleic acid, protein, and other types.

(1) Nucleic acid research

Nucleic acid research includes sequencing and application, gene sequence annotation, gene prediction, nucleotide sequence alignment, nucleic acid database, comparative genomics, metagenomics, gene evolution, and RNA structure prediction.

(2) Protein research

Protein research includes protein database, protein sequence alignment, prediction of protein secondary and tertiary structure, protein molecular interaction, molecular docking, proteomics, etc.

(3) Other contents

There are other contents such as metabolic networks, big data mining, algorithm development, computational evolutionary biology, biodiversity and so on.

## III. Common database of bioinformatics

Bioinformatics databases can be divided into three main categories: nucleic acid database, protein database, and special database.

(1) Nucleic acid database

GenBank and the DNA database of the European Molecular Biology Laboratory( EMBL) and the DNA database in Japan are the three largest DNA databases in the world, which together constitute the international database of nucleic acid sequences. The three organizations exchange and update data and information every day.

1) GenBank: belonging to the National Center for Biotechnology Information ( NCBI) of the United States, is the largest database to provide all the public DNA sequences. The nucleic acid sequences of Gen-Bank can be accessed openly, each record is annotated with the coding region ( CDS) , and also the translation of amino acids.

2) EBI: DNA database of The European Molecular Biology Laboratory ( EMBL) .

3) DDBJ ( DNA Data Bank of Japan) , the Japanese DNA database, was established in 1984. The information is exchanged in two international annual conferences including the international DNA database consultancy conference and the international DNA database collaboration meeting, so the data of the three libraries are actually the same.

(2) Protein database

1) Uniprot ( http://www. uniprot. org) : Uniprot, the abbreviation of Universal Protein, is the protein database with the richest information and the most extensive resources. It is made up of data from three major databases: Swiss−Prot, TrEMBL, and PIR−PSD. Its data mainly came from the protein sequences after the completion of the genome sequencing project.

2) PDB ( Protein Data Bank) : PDB is the only database in the world that stores the 3D structure of biological macromolecules. In addition to protein, these biological macromolecules include nucleic acids, nucleic acid and protein complexes, and only three stages of the structure obtained through experimental methods will be recorded. PDB was first created in 1971 by the Brookhaven National Laboratory. At present, the PDB database is updated once a week. So far, more than 100 thousand structures have been collected, of which more than 90% are protein structures.

(3) Special database.

1) KEGG ( Kyoto Encyclopedia of Genes and Genomes) : the Encyclopedia of gene and genome of Kyoto, which was established in 1995 by Kanehisa Laboratory of the center for bioinformatics at Kyoto University in Japan. It is a database for systematic analysis of gene function. It connects genome information with gene function to reveal the mystery of life.

2) OMIM ( Online Mendelian Inheritance in Man) : "human Mendel genetic online" is a database with respect to human genes and genetic disorders. It mainly focuses on genetic or genetic diseases, including text information and related reference information, sequence records, maps, and other related databases.

3) Genecards: the human gene database. When studying a disease−related gene, this database can provide a number of related information related to the gene, including expression, function, genetic information, pathway, protein and so on.

# Experiment 1　DNA Sequence Similarity Search and Primer Design by Using BLAST

## 1.1　Principle

Basic Local Alignment Search Tool( BLAST) is a set of analytical tools for similarity comparison in protein databases or DNA databases. The BLAST tool is selected according to the type of query sequence ( protein or nucleic acid). There are five tools of BLAST.

BLASTP is a kind of query in protein sequence to protein library. Each known sequence in the library will be compared one by one with each checked sequence one by one.

BLASTX is a kind of query in nucleic acid sequence to protein library. The sequence of nucleic acid is first translated into protein sequence (the sequence of nucleic acids will be translated into six possible proteins), and one to one protein sequence alignment is made.

BLASTN is a query of nucleic acid sequence to nucleic acid library. Each known sequence in the library will be compared to the query nucleic acid sequence one by one.

TBLASTN is a kind of query in protein sequence to nucleic acid library. Contrary to BLASTX, it translates the nucleic acid sequence in the library into protein sequence and then compares this protein sequence with the query protein sequence.

TBLASTX is a kind of query in nucleic acid sequence to nucleic acid library.

Primer—BLAST is a very useful PCR primer design tool by using BLAST. It only needs to submit the template sequence, and then the available primers will be automatically designed. Because it is an online search combined with sequence alignment, the primer specificity can be ensured.

## 1.2　Materials

Computers with Internet connections and web browsers; DNA sequences.

## 1.3　Methods and results

### 1.3.1　Comparison of p53 gene by using BLAST

Log onto the NCBI home page (https://www.ncbi.nlm.nih.gov/), click"BLAST" on the right of the screen (under "Popular Resources") (Figure 24.1A), enter the interface (https://blast.ncbi.nlm.nih.gov/Blast.cgi), select the " Nucleotide BLAST " method (Figure 24.1B).

A.

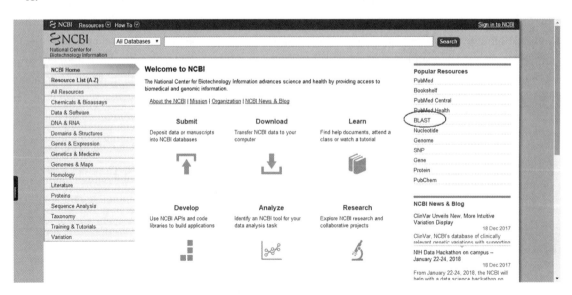

B.

Web BLAST

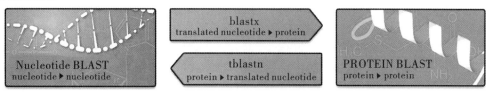

Figure 24.1   **Home page**
A. NCBI home page; B. BLAST home page.

(1) BLAST analysis of human p53 gene sequence

There are two ways of submitting the sequence. One is to copy the DNA sequence directly ( with the human p53 gene sequence as an example) and paste into "Enter Query Sequence" box; the other is to use the browses button to upload a file from local disk, the file may contain a single sequence or a list of sequences ( Figure 24.2A), and click "BLAST" ( Figure 24.2B).

A.

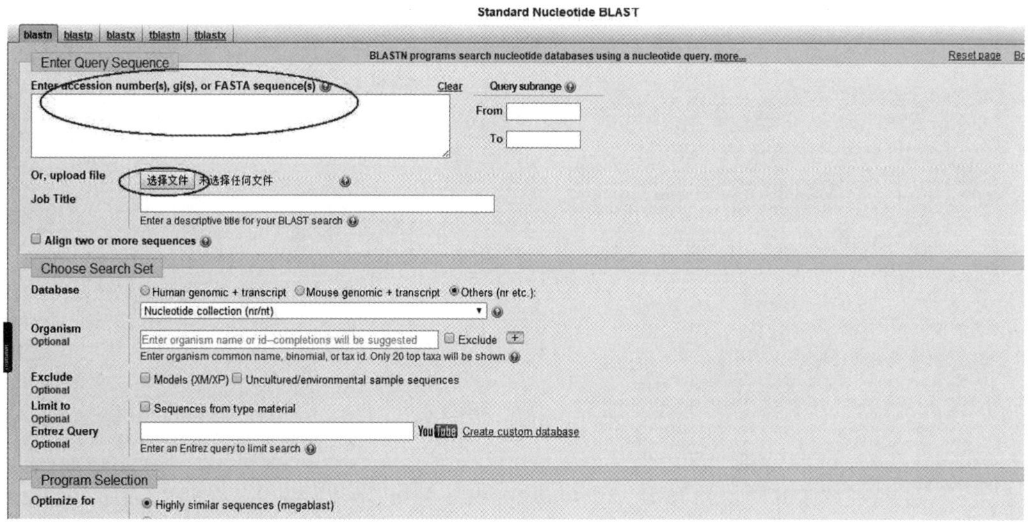

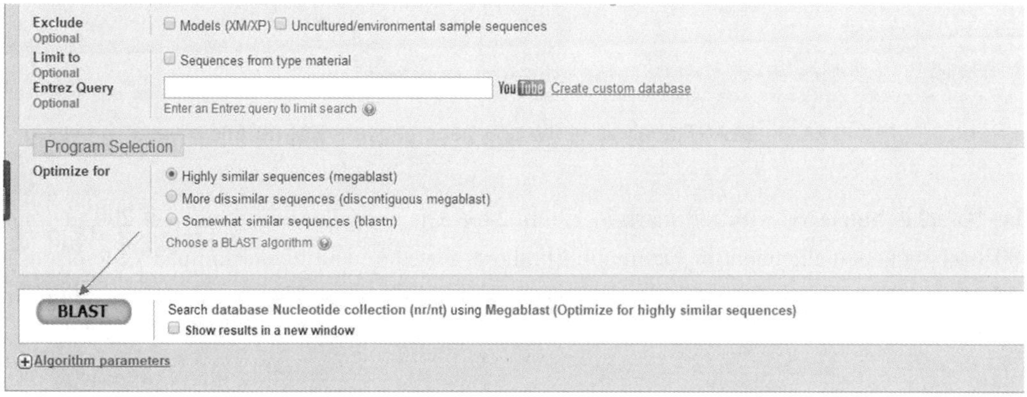

Figure 24.2   **Two ways of submitting the sequence**

(2) BLAST analysis of the p53 gene sequence of rat and mouse

Use "Align two or multiple sequences" button ( Figure 24.3A), enter one query in the top text box ( the p53 gene sequences of rat) and one subject sequence in the lower box ( the p53 gene sequences of mouse), and then click the BLAST button at the bottom of the page to align the sequences ( Figure 24.3B).

A.

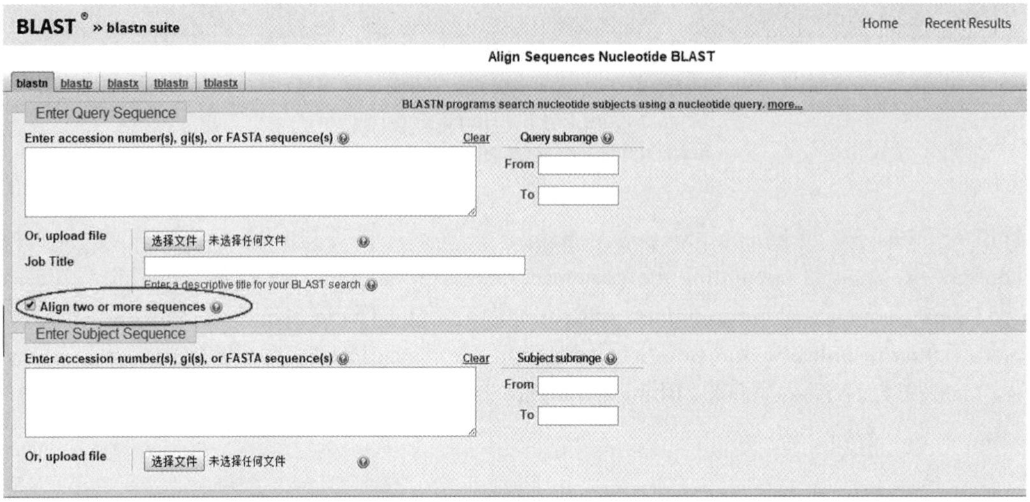

B.

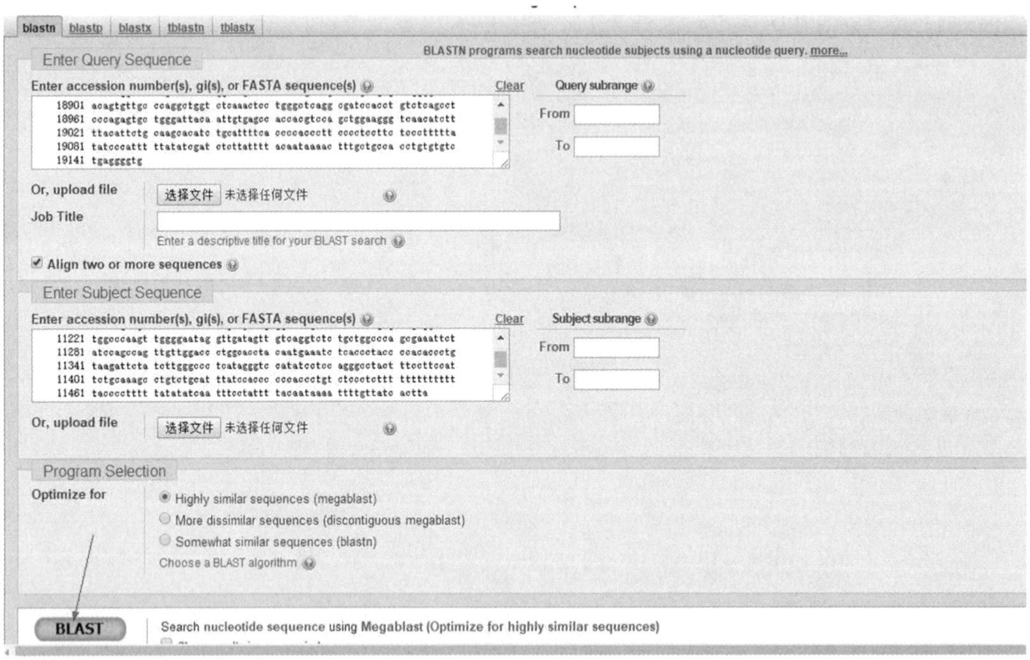

Figure 24.3   BLAST analysis of the p53 gene sequence of rat and mouse

The "Graphic Summary" with red mark in Figure 24.4A is a section with a score of 200 and more; the 1−1,500 base sequence alignment in Figure 24.4B shows that the identities of the p53 gene of rat and the mouse p53 gene are 1,302/1,500 (87%). The results showed that the p53 gene of rat was quite similar to that of p53 gene in mouse.

A.

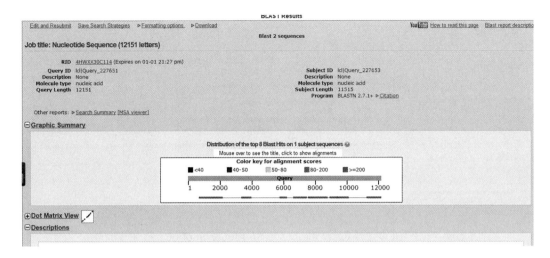

B.

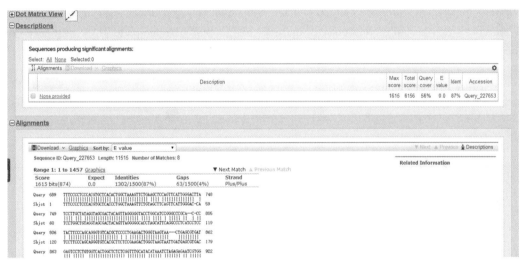

Figure 24. 4   BLAST result of the p53 gene sequence of rat and mouse

(3) BLAST analysis of the p53 gene sequence of human and mouse

Repeat the procedure of step 2, and submit the p53 gene sequence of human and mouse into the top and lower boxes. The result of BLAST shows that the similarity between the human p53 gene and mouse p53 gene is very low (Figure 24. 5).

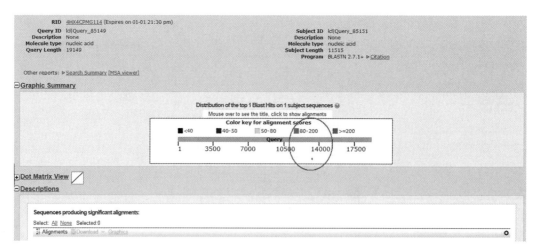

Figure 24.5   BLAST result of the p53 gene sequence of human and mouse

Design PCR primers with Primer–BLAST.

Log onto the BLAST interface ( https://blast. ncbi. nlm. nih. gov/Blast. cgi) and select "Primer–BLAST searches" in Specialized searches ( Figure 24. 6), submit the template DNA sequence, and the available primers will be automatically designed.

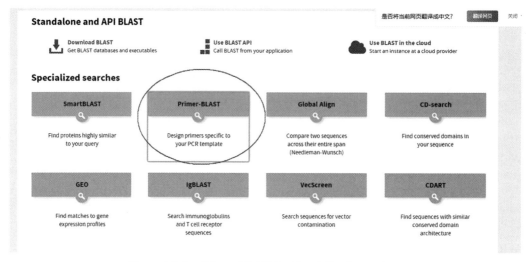

Figure 24.6   Primer–BLAST in Specialized searches

Input PCR template DNA sequence, set the relevant parameters of PCR primers ( follow the principle of primer design, see Chapter 20), submit ( Figure 24. 7).

Figure 24.7    **Prime–BLAST homepage**

The available primers and details of the system are shown in Figure 24. 8.

A.

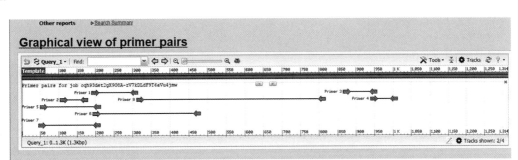

B.

是否将当前网页翻译成中文？　翻译网页

**Primer pair 1**

| | Sequence (5'->3') | Template strand | Length | Start | Stop | Tm | GC% | Self complementarity | Self 3' complementarity |
|---|---|---|---|---|---|---|---|---|---|
| Forward primer | AGGGCTCACTCCAGGTAAGT | Plus | 20 | 190 | 209 | 59.88 | 55.00 | 3.00 | 1.00 |
| Reverse primer | TGCAGCCCTAAGCATCTAGC | Minus | 20 | 314 | 295 | 59.89 | 55.00 | 4.00 | 2.00 |
| Product length | 125 | | | | | | | | |

**Primer pair 2**

| | Sequence (5'->3') | Template strand | Length | Start | Stop | Tm | GC% | Self complementarity | Self 3' complementarity |
|---|---|---|---|---|---|---|---|---|---|
| Forward primer | GGCGTAAACGCTTCGAGATG | Plus | 20 | 104 | 123 | 59.70 | 55.00 | 6.00 | 2.00 |
| Reverse primer | GACTCCTCTGTAGCATGGGC | Minus | 20 | 179 | 160 | 59.89 | 60.00 | 4.00 | 2.00 |
| Product length | 76 | | | | | | | | |

**Primer pair 3**

| | Sequence (5'->3') | Template strand | Length | Start | Stop | Tm | GC% | Self complementarity | Self 3' complementarity |
|---|---|---|---|---|---|---|---|---|---|
| Forward primer | CAGACTGACTGCCTCTGCAT | Plus | 20 | 865 | 884 | 59.75 | 55.00 | 5.00 | 2.00 |
| Reverse primer | CCGGGAGGATTGTGTCTCAG | Minus | 20 | 956 | 937 | 59.82 | 60.00 | 4.00 | 3.00 |
| Product length | 92 | | | | | | | | |

**Primer pair 4**

| | Sequence (5'->3') | Template strand | Length | Start | Stop | Tm | GC% | Self complementarity | Self 3' complementarity |
|---|---|---|---|---|---|---|---|---|---|
| Forward primer | GAGACACAATCCTCCCGGTC | Plus | 20 | 939 | 958 | 59.82 | 60.00 | 4.00 | 1.00 |
| Reverse primer | TACTCAGAGAGGGGGCTGAG | Minus | 20 | 1011 | 992 | 59.74 | 60.00 | 5.00 | 3.00 |
| Product length | 73 | | | | | | | | |

**Primer pair 5**

| | Sequence (5'->3') | Template strand | Length | Start | Stop | Tm | GC% | Self complementarity | Self 3' complementarity |
|---|---|---|---|---|---|---|---|---|---|
| Forward primer | TGTGTGACCTTGTCCAGTGC | Plus | 20 | 50 | 69 | 60.46 | 55.00 | 3.00 | 2.00 |
| Reverse primer | AGGCCACTTACCTGGAGTGA | Minus | 20 | 214 | 195 | 60.18 | 55.00 | 4.00 | 1.00 |
| Product length | 165 | | | | | | | | |

**Primer pair 6**

Figure 24.8　Prime–BLAST Result

# 1.4　Discussion

(1) Compare the similarity of p53 gene between human p53 gene and mouse or rat.

(2) The relation of the sequences difference to the evolution time of species.

*Zhang Xu, Ji Yupei*

# Appendix A

# Preparation and Storage of Chemical Reagents

## A-1 The Purity Grade and Storage of Chemical Reagents

## I. The purity grade of the chemical reagents

The purity grade of chemical reagents are shown in Table A. 1.

Table A. 1　The purity grade of chemical reagents

|  | Grade one | Grade two | Grade three | Grade four | Biological reagents |
|---|---|---|---|---|---|
| Domestic standards | Guaranteed reagents, G. R., with a green label on the reagent bottle | Analytical pure reagents, A. R., with a red label on the reagent bottle | Chemical pure reagents, C. P., with a blue label on the reagent bottle | Experimental pure reagents, L. R., with a yellow label on the reagent bottle | B. R. or C. R. |
| International standards | A. R.<br>G. R.<br>A. C. S.<br>P. A. | C. P.<br>P. U. S. S.<br>PURISS | L. A. P.<br>E. P. | X | |
| The application for chemical reagents | Highpurity, suitable for research work and the precise analysis, especially for preparing a standard solution | High purity, suitable for chemical qualitative and quantitative analysis | High purity, slightly lower than that of grade two, its application is similar to that of grade two | | Follow the manufacturer's instructions |

# II. General instructions for preparation of chemical reagents

(1) Weigh chemical reagents accurately, especially when preparing a standard solution or buffer. In some special cases, the dehydration, constant weight and purification for some chemical reagents should be considered.

(2) Unless the special circumstances, most of the reagent solution should be made up with distilled water or deionized water (ion-exchange water).

(3) Chemical reagents are usually divided into various grades (specifications) according to their purity. Besides purity grades, there are other grades such as spectrophotometric grade with high purity, chromatographic and pharmaceutical grade with relative low purity.

(4) Chemical reagent solution should be prepared according to the minimal volume required. Too much is a waste.

(5) If chemical reagents (especially liquid) are taken out from reagent bottles, all the remaining mustn't be put back to the original bottle. Because the contaminated beaker or spoon could cause the cross-contamination, it is required to ensure reagent spoons clean and dry.

(6) All of the glassware and bottles used for preparing the chemical solutions and storing the reagents should be carefully cleaned.

(7) The chemical solution should be stored and labeled with the name and concentration of reagents, preparation date, and preparer.

(8) Glass bottles contained chemical solution should be plugged tightly with clean bottle plugs. For safe storage of chemicals, the plugs mustn't contact with contaminated items.

(9) Some chemical reagents are prone to deteriorate with time, so pay more attention to the expiration date.

# III. Safety storage for some special chemical reagents

The Safety storage for some special chemical reagents are shown in Table A.2.

Table A.2 Safety storage for some special chemical reagents

| Safety storage | Notices | Chemical names |
|---|---|---|
| Stored in tight container | Easy deliquescence reagents | Calcium oxide, sodium hydroxide, potassium hydroxide, potassium iodideand trichloroacetic acid |
| | Easy dehydration reagents | Sodium sulfate anhydrous, ferrous sulfate, sodium hydrogen phosphateand sodium thiosulfate |
| | Easy volatilization reagents | Ammonium hydroxide, chloroform, ether, iodine, thymol crystals, formaldehyde, ethyl alcohol and acetone |
| | Carbon dioxide absorbed reagents | Potassium hydroxide and sodium hydroxide |
| | Easily oxidized reagents | Ferrous sulfate, ether, aldehydes, phenol, acid and all reductants |
| | Perishable reagents | Sodium pyruvate, diethyl ether and many biological products (kept in cold storage) |

Continue to Table A.2

| Safety storage | Notices | Chemical names |
|---|---|---|
| Stored in amber bottle | Photochromic reagents | Silver nitrate ( blackening), phenol ( variable pink), chloroform ( producing phosgene) and ninhydrin ( variable pink) |
| | Photolysis reagents | Hydrogen peroxide, chloroform, bleaching powder and hydrogen cyanide |
| | Photooxidation reagents | Ether, aldehyde, ferrous salt and all reductants |
| Stored in special ways | Explosive reagents | Picric acid, nitrates, perchloric acid and sodium chloride |
| | Rank poison | Potassium cyanide, sodium cyanide, mercury, arsenide and bromine |
| | Inflammable reagents | Bromide ether, methanol, ethanol, propyl alcohol, benzene, methylbenzene, dimethylbenzene and gasoline |
| | Corrosive reagents | strong acid and alkali |

Some chemical reagents need to be stored in sealed reagent bottles, tightly sealed with wax. Besides, some chemicals should be stored in the desiccator whose desiccants could be lime, anhydrous sodium chlorideand silica gel. Although concentrated sulfuric acid has a capacity to absorb water, it can not be used as a desiccant. Additionally, light-sensitive chemical reagents should be placed in brown bottles or wrapped bottles by black papers.

# A-2   Preparation of Common Chemical Solution as Well as Buffer Solution

## I . Preparation of common chemical solution

(1)0.5 mol/L EDTA ( pH 8.0)

Weigh 93.06 g $Na_2EDTA \cdot 2H_2O$ to 500 mL beaker, add 400 mL $ddH_2O$ and stir the solution vigorously using a thermal magnetic stirrer. Adjust pH value to 8.0 by adding solid NaOH pellets ( -10 g), and make up the final volume to 500 mL with $ddH_2O$. After sterilization by autoclave, store at room temperature.

(2) 10 mol/L ammonium acetate solution

Dissolve 77.1 g ammonium acetate in 30 mL $ddH_2O$ in 100-200 mL beaker and then make up the final volume to 1 L with $ddH_2O$. Filter through 0.22 μm filter and store in a sealed bottle at room temperature.

*Notes: Ammonium acetate is easily decomposed if heated, so high pressure-high temperature sterilization mustn't be used for it.*

(3)2 mol/L NaOH solution

Weigh 20 g NaOH, add $ddH_2O$ to the final volume 100 mL. Store at room temperature.

(4)2.5 mol/L HCl solution

Add 21.6 mL of concentrated HCl (11.6 mol/L) in 78.4 mL $ddH_2O$ and mix fully. Store at room temperature.

(5)5 mol/L NaCl solution

Dissolve 292.2 g NaCl in -800 mL $ddH_2O$ in 1 L beaker and mix fully, make up the final volume to

1 L with ddH$_2$O. After sterilization by autoclave, aliquot and store at room temperature.

(6) 20% (W/V) glucose solution

Dissolve 20 g glucose in −80 mL ddH$_2$O in 100−200 mL beaker and mix fully, make up the final volume to 100 mL with ddH$_2$O. After sterilization by autoclave, store at room temperature.

(7) Diethyl pyrocarbonate (DEPC) −treated water

Add 100 μL DEPC in 100 mL ddH$_2$O (0.1% DEPC), and then incubate DEPC solution at 37 ℃ for 12 hours in a water bath. Autoclaved (at least 20 min) under 15 psi condition to inactivate the residual DEPC. DEPC water mustn't be used to prepare Tris—based buffer, because DEPC reacts with a free amino group, Tris buffer will lose its buffer capacity.

(8) TE buffer (for suspension and storage of DNA)

1) Buffer recipes: 10 mmol/L Tris−HCl, 1 mmol/L EDTA.

2) Chemical amount in 100 mL TE buffer: 1 mL ddH$_2$O, 200 μL 1 mol/L Tris−HCl(pH 7.4− 8.0 25 ℃), 98.8 mL 0.5 mol/L EDTA (PH 8.0).

(9) PBS buffer (137 mmol/L NaCl, 2.7mmol/L KCl, 10 mmol/L Na$_2$HPO$_4$, 2 mmol/L KH$_2$PO$_4$)

Dissolve 8 g NaCl, 0.2 g KCl, 41.42 g Na$_2$HPO$_4$ and 40.27 g KH$_2$PO$_4$ in 800 mL ddH$_2$O in 1 L beaker. Adjust pH value to 7.4 with HCl solution, and make up the final volume to 1 L with ddH$_2$O. After sterilization by autoclave, store at room temperature.

*Notes: There are no bivalent cation in above PBS buffer, if needed, supplement 1 mmol/L CaCl$_2$ or 0.5 mmol/L MgCl$_2$ in the above recipe.*

# II. Preparation of common buffer solution

(1) Disodium hydrogen phosphate−citric acid buffer

The relative molecular weight of Na$_2$HPO$_4$ is 141.98, 0.2 molar of Na$_2$HPO$_4$ contains 28.40 g/L per liter. The relative molecular weight of Na$_2$HPO$_4$ · 2H$_2$O is 178.05, 0.2 molar of Na$_2$HPO$_4$ · 2H$_2$O contains 35.61 g/L. The relative molecular weight of C$_6$H$_8$O$_7$ · H$_2$O is 210.14, 0.1 molar of C$_6$H$_8$O$_7$ · H$_2$O contains 21.01 g/L (Table A.3).

Table A.3　Disodium hydrogen phosphate−citric acid buffer

| pH | 0.2 mol/L disodium hydrogen phosphate(mL) | 0.1 mol/L citric acid(mL) | pH | 0.2 mol/L disodium hydrogen phosphate (mL) | 0.1 mol/L citric acid (mL) |
|---|---|---|---|---|---|
| 2.2 | 0.40 | 19.60 | 5.2 | 10.72 | 9.28 |
| 2.4 | 1.24 | 18.76 | 5.4 | 11.15 | 8.85 |
| 2.6 | 2.18 | 17.82 | 5.6 | 11.60 | 8.40 |
| 2.8 | 3.17 | 16.83 | 5.8 | 12.09 | 7.91 |
| 3.0 | 4.11 | 15.89 | 6.0 | 12.63 | 7.37 |
| 3.2 | 4.94 | 15.06 | 6.2 | 13.22 | 6.78 |
| 3.4 | 5.70 | 14.30 | 6.4 | 13.85 | 6.15 |
| 3.6 | 6.44 | 13.56 | 6.6 | 14.55 | 5.45 |
| 3.8 | 7.10 | 12.90 | 6.8 | 15.45 | 4.55 |
| 4.0 | 7.71 | 12.29 | 7.0 | 16.47 | 3.53 |

Continue to Table A.3

| pH | 0.2 mol/L disodium hydrogen phosphate(mL) | 0.1 mol/L citric acid(mL) | pH | 0.2 mol/L disodium hydrogen phosphate (mL) | 0.1 mol/L citric acid (mL) |
|---|---|---|---|---|---|
| 4.2 | 8.28 | 11.72 | 7.2 | 17.39 | 2.61 |
| 4.4 | 8.82 | 11.18 | 7.4 | 18.17 | 1.83 |
| 4.6 | 9.35 | 10.65 | 7.6 | 18.73 | 1.27 |
| 4.8 | 9.86 | 10.14 | 7.8 | 19.15 | 0.83 |
| 5.0 | 10.30 | 9.70 | 8.0 | 19.45 | 0.55 |

(2) Citric acid–sodium hydroxide–hydrochloric acid buffer

If needed, add 1 g phenol per liter. If pH value has been changed, adjust it by adding a small amount of 12.5 mol/L (50%) NaOH or concentrated hydrochloric acid. Store at 4 ℃ in fridge(Table A.4).

Table A.4   Citric acid–sodium hydroxide–hydrochloric acid buffer

| pH | Concentration of sodium ion (mol/L) | Citric acid (g) $C_6H_8O_7 \cdot H_2O$ | Sodium hydroxide (g) NaOH (97%) | Hydrochloric acid (mL) HCl (strong) | Final volume (L) |
|---|---|---|---|---|---|
| 2.2 | 0.20 | 210 | 84 | 160 | 10 |
| 3.1 | 0.20 | 210 | 83 | 116 | 10 |
| 3.3 | 0.20 | 210 | 83 | 106 | 10 |
| 4.3 | 0.20 | 210 | 83 | 45 | 10 |
| 5.3 | 0.35 | 245 | 144 | 68 | 10 |
| 5.8 | 0.45 | 285 | 186 | 105 | 10 |
| 6.5 | 0.38 | 266 | 156 | 126 | 10 |

(3) Citric acid–sodium citrate buffer (0.1 mol/L)

The relative molecular weight of $C_6H_8O_7 \cdot H_2O$ is 210.14, 0.1 molar of $C_6H_8O_7 \cdot H_2O$ contains 21.01 g/L. The relative molecular weight of $Na_3C_6H_5O_7 \cdot 2H_2O$ is 294.12, 0.1 molar of $Na_3C_6H_5O_7 \cdot 2H_2O$ contains 29.41 g/L(Table A.5).

Table A.5   Citric acid–sodium citrate buffer (0.1 mol/L)

| pH | 0.1 mol/L citric acid (mL) | 0.1 mol/L sodium citrate (mL) | pH | 0.1 mol/L citric acid (mL) | 0.1 mol/L sodium citrate (mL) |
|---|---|---|---|---|---|
| 3.0 | 18.6 | 1.4 | 5.0 | 8.2 | 11.8 |
| 3.2 | 17.2 | 2.8 | 5.2 | 7.3 | 12.7 |
| 3.4 | 16.0 | 4.0 | 5.4 | 6.4 | 13.6 |
| 3.6 | 14.9 | 5.1 | 5.6 | 5.5 | 14.5 |
| 3.8 | 14.0 | 6.0 | 5.8 | 4.7 | 15.3 |
| 4.0 | 13.1 | 6.9 | 6.0 | 3.8 | 16.2 |
| 4.2 | 12.3 | 7.7 | 6.2 | 2.8 | 17.2 |

Continue to Table A.5

| pH | 0.1 mol/L citric acid (mL) | 0.1 mol/L sodium citrate (mL) | pH | 0.1 mol/L citric acid (mL) | 0.1 mol/L sodium citrate (mL) |
|----|----|----|----|----|----|
| 4.4 | 11.4 | 8.6 | 6.4 | 2.0 | 18.0 |
| 4.6 | 10.3 | 9.7 | 6.6 | 1.4 | 18.6 |
| 4.8 | 9.2 | 10.8 | | | |

(4) Phosphate buffered saline

1) Disodium hydrogen phosphate–sodium dihydrogen phosphate buffer (0.2 mol/L)

The relative molecular weight of $Na_2HPO_4 \cdot 2H_2O$ is 178.05, 0.2 molar of $Na_2HPO_4 \cdot 2H_2O$ contains 35.61 g/L. The relative molecular weight of $Na_2HPO_4 \cdot 12H_2O$ is 358.22, 0.2 molar of $Na_2HPO_4 \cdot 12H_2O$ contains 71.64 g/L. The relative molecular weight of $NaH_2PO_4 \cdot H_2O$ is 138.01, 0.2 molar of $NaH_2PO_4 \cdot H_2O$ contains 27.6 g/L. The relative molecular weight of $NaH_2PO_4 \cdot 2H_2O$ is 156.03, 0.2 molar of $NaH_2PO_4 \cdot 2H_2O$ contains 31.21 g/L(Table A.6).

Table A.6　Disodium hydrogen phosphate–sodium dihydrogen phosphate buffer (0.2 mol/L)

| pH | 0.2 mol/L $Na_2HPO_4$(mL) | 0.2 mol/L $NaH_2PO_4$(mL) | pH | 0.2 mol/L $Na_2HPO_4$(mL) | 0.2 mol/L $NaH_2PO_4$(mL) |
|----|----|----|----|----|----|
| 5.8 | 8.0 | 92.0 | 7.0 | 61.0 | 39.0 |
| 5.9 | 10.0 | 90.0 | 7.1 | 67.0 | 33.0 |
| 6.0 | 12.3 | 87.7 | 7.2 | 72.0 | 28.0 |
| 6.1 | 15.0 | 85.0 | 7.3 | 77.0 | 23.0 |
| 6.2 | 18.5 | 81.5 | 7.4 | 81.0 | 19.0 |
| 6.3 | 22.5 | 77.5 | 7.5 | 84.0 | 16.0 |
| 6.4 | 26.5 | 73.5 | 7.6 | 87.0 | 13.0 |
| 6.5 | 31.5 | 68.5 | 7.7 | 89.5 | 10.5 |
| 6.6 | 37.5 | 62.5 | 7.8 | 91.5 | 8.5 |
| 6.7 | 43.5 | 56.5 | 7.9 | 93.0 | 7.0 |
| 6.8 | 49.0 | 51.0 | 8.0 | 94.7 | 5.3 |
| 6.9 | 55.0 | 45.0 | | | |

2) Disodium hydrogen phosphate–potassium dihydrogen phosphate buffer (1/15 mol/L)

The relative molecular weight of $Na_2HPO_4 \cdot 2H_2O$ is 178.05, 1/15 molar of $Na_2HPO_4 \cdot 2H_2O$ contains 11.876 g/L. The relative molecular weight of $KH_2PO_4$ is 136.09, 1/15 molar of $KH_2PO_4$ contains 9.078 g/L(Table A.7).

Table A.7  Disodium hydrogen phosphate–sodium dihydrogen phosphate buffer (1/15 mol/L)

| pH | 1/15 mol/L Na$_2$HPO$_4$(mL) | 1/15 mol/L KH$_2$PO$_4$(mL) | pH | 1/15 mol/L Na$_2$HPO$_4$(mL) | 1/15 mol/L KH$_2$PO$_4$(mL) |
|---|---|---|---|---|---|
| 4.92 | 0.10 | 9.90 | 7.17 | 7.00 | 3.00 |
| 5.29 | 0.50 | 9.50 | 7.38 | 8.00 | 2.00 |
| 5.91 | 1.00 | 9.00 | 7.73 | 9.00 | 1.00 |
| 6.24 | 2.00 | 8.00 | 8.04 | 9.50 | 0.50 |
| 6.47 | 3.00 | 7.00 | 8.34 | 9.75 | 0.25 |
| 6.64 | 4.00 | 6.00 | 8.67 | 9.90 | 0.10 |
| 6.81 | 5.00 | 5.00 | 8.18 | 10.00 | 0.00 |
| 6.98 | 6.00 | 4.00 | | | |

(5) Potassium dihydrogen phosphate–sodium hydrate buffer (0.05 mol/L)

$x$ mL 0.2 mol/L KH$_2$PO$_4$ + $y$ mL 0.2 mol/L NaOH, add ddH$_2$O to 20 mL(Table A.8).

Table A.8  Potassium dihydrogen phosphate–sodium hydrate buffer (0.05 mol/L)

| pH(20 ℃) | $x$ | $y$ | pH(20 ℃) | $x$ | $y$ |
|---|---|---|---|---|---|
| 5.8 | 5 | 0.372 | 7.0 | 5 | 2.963 |
| 6.0 | 5 | 0.570 | 7.2 | 5 | 3.500 |
| 6.2 | 5 | 0.860 | 7.4 | 5 | 3.950 |
| 6.4 | 5 | 1.260 | 7.6 | 5 | 4.280 |
| 6.6 | 5 | 1.780 | 7.8 | 5 | 4.520 |
| 6.8 | 5 | 2.365 | 8.0 | 5 | 4.680 |

(6) Barbiturate sodium–hydrochloric acid buffer (18 ℃)

The relative molecular weight of barbiturate sodium is 206.18, 0.04 molar of barbiturate sodium contains 8.25 g/L(Table A.9).

Table A.9  Barbiturate sodium–hydrochloric acid buffer (18 ℃)

| pH | 0.04 mol/L barbiturate sodium (mL) | 0.2 mol/L hydrochloric acid (mL) | pH | 0.04 mol/L barbiturate sodium(mL) | 0.2 mol/L hydrochloric acid (mL) |
|---|---|---|---|---|---|
| 6.8 | 100 | 18.40 | 8.4 | 100 | 5.21 |
| 7.0 | 100 | 17.80 | 8.6 | 100 | 3.82 |
| 7.2 | 100 | 16.70 | 8.8 | 100 | 2.52 |
| 7.4 | 100 | 15.30 | 9.0 | 100 | 1.65 |
| 7.6 | 100 | 13.40 | 9.2 | 100 | 1.13 |
| 7.8 | 100 | 11.47 | 9.4 | 100 | 0.70 |
| 8.0 | 100 | 9.39 | 9.6 | 100 | 0.35 |
| 8.2 | 100 | 7.21 | | | |

(7) Trihydroxymethyl aminomethane( Tris) —hydrochloric acid buffer( 25 ℃)

50 mL 0.1 mol/L trihydroxymethyl aminomethane ( Tris) + $x$ mL 0.1 mol/L hydrochloric acid, add ddH$_2$O to 100 mL. The relative molecular weight of $H_2N—C(CO_2OH)_3$ is 121.14, 0.1 molar of $H_2N—C(CO_2OH)_3$ contains 12.114 g/L. Tris buffer is usually stored in a closed container because it will absorb carbon dioxide from the air( Table A. 10).

Table A. 10   Trihydroxymethyl aminomethane–hydrochloric acid buffer(25 ℃)

| pH | $x$ | pH | $x$ |
|----|----|----|----|
| 7.1 | 45.7 | 8.1 | 26.2 |
| 7.2 | 44.7 | 8.2 | 22.9 |
| 7.3 | 43.4 | 8.3 | 19.9 |
| 7.4 | 42.0 | 8.4 | 17.2 |
| 7.5 | 40.3 | 8.5 | 14.7 |
| 7.6 | 38.5 | 8.6 | 12.4 |
| 7.7 | 36.6 | 8.7 | 10.3 |
| 7.8 | 34.5 | 8.8 | 8.5 |
| 7.9 | 32.0 | 8.9 | 7.0 |
| 8.0 | 29.2 | 9.0 | 5.7 |

(8) 10×TE buffer( Table A. 11).

Table A. 11   10×TE buffer

| 10×TE buffer ( different pH) | The volume of stock solution( mL) |
|----|----|
| pH 7.4 | |
| 1 mol/L Tris–HCl ( pH 7.4) | 100 mL |
| 500 mmol/L EDTA ( pH 8.0) | 20 mL |
| | add ddH$_2$O to 1,000 mL |
| pH 7.6 | |
| 1 mol/L Tris–HCl ( pH 7.6) | 100 mL |
| 500 mmol/L EDTA ( pH 8.0) | 20 mL |
| | add ddH$_2$O to 1,000 mL |
| pH 8.0 | |
| 1 mol/L Tris–HCl ( pH 8.0) | 100 mL |
| 500 mmol/L EDTA ( pH 8.0) | 20 mL |
| | add ddH$_2$O to 1,000 mL |

The above buffer needs to be sterilized by autoclave, store at room temperature.

(9) Boric acid–borax buffer

The relative molecular weight of $Na_2B_4O_7 \cdot 10H_2O$ is 381.43, 0.05 molar of $Na_2B_4O_7 \cdot 10H_2O$ contains 19.07 g/L; the relative molecular weight of $H_3BO_3$ is 61.84, 0.2 molar of $H_3BO_3$ contains 12.37 g/L. Borax is easier to lose crystal water, so it must be stored in a capped reagent bottle( Table A. 12).

Table A.12  Boric acid–borax buffer

| pH | 0.05 mol/L borax(mL) | 0.2 mol/L boric acid(mL) | pH | 0.05 mol/L borax(mL) | 0.2 mol/L boric acid(mL) |
|---|---|---|---|---|---|
| 7.4 | 1.0 | 9.0 | 8.2 | 3.5 | 6.5 |
| 7.6 | 1.5 | 8.5 | 8.4 | 4.5 | 5.5 |
| 7.8 | 2.0 | 8.0 | 8.7 | 6.0 | 4.0 |
| 8.0 | 3.0 | 7.0 | 9.0 | 8.0 | 2.0 |

(10) Glycine–sodium hydrate buffer (0.05 mol/L)

$x$ mL of 0.2 mol/L glycine + $y$ mL of 0.2 mol/L NaOH, dilute buffer to 200 mL. The relative molecular weight of glycine is 75.07, 0.2 molar of glycine contains 15.01 g/L(Table A.13).

Table A.13  Glycine–sodium hydrate buffer (0.05 mol/L)

| pH | $x$ | $y$ | pH | $x$ | $y$ |
|---|---|---|---|---|---|
| 8.6 | 50 | 4.0 | 9.6 | 50 | 22.4 |
| 8.8 | 50 | 6.0 | 9.8 | 50 | 27.2 |
| 9.0 | 50 | 8.8 | 10.0 | 50 | 32.0 |
| 9.2 | 50 | 12.0 | 10.4 | 50 | 38.6 |
| 9.4 | 50 | 16.8 | 10.6 | 50 | 45.5 |

(11) Borax–sodium hydrate buffer (0.05 mol/L boric acid root)

$x$ mL of 0.05 mol/L borax + $y$ mL of 0.2 mol/L NaOH, dilute buffer to 200 mL. The relative molecular weight of $Na_2B_4O_7 \cdot 10H_2O$ is 381.43, 0.05 molar of boric acid root solution (0.2 mol/L borax) contains 19.07 g/L(Table A.14).

Table A.14  Borax–sodium hydrate buffer (0.05 mol/L boric acid root)

| pH | $x$ | $y$ | pH | $x$ | $y$ |
|---|---|---|---|---|---|
| 9.3 | 50 | 6.0 | 9.8 | 50 | 34.0 |
| 9.4 | 50 | 11.0 | 10.0 | 50 | 43.0 |
| 9.6 | 50 | 23.0 | 10.1 | 50 | 46.0 |

(12) Sodium carbonate–sodium bicarbonate buffer (0.1 mol/L)

The relative molecular weight of $Na_2CO_3 \cdot 10H_2O$ is 286.2, 0.1 molar of $Na_2CO_3 \cdot 10H_2O$ contains 8.62 g/L. The relative molecular weight of $NaHCO_3$ is 84.0, 0.1 molar of $NaHCO_3$ contains 8.40 g/L. This buffer cannot be used in the test when $Ca^{2+}$, $Mg^{2+}$ exist(Table A.15).

Table A.15  Sodium carbonate-sodium bicarbonate buffer (0.1 mol/L)

| pH | | 0.1 mol/L | 0.1 mol/L |
|---|---|---|---|
| 20 ℃ | 37 ℃ | $Na_2CO_3$(mL) | $NaHCO_3$(mL) |
| 9.16 | 8.77 | 1 | 9 |
| 9.40 | 9.12 | 2 | 8 |
| 9.51 | 9.40 | 3 | 7 |
| 9.78 | 9.50 | 4 | 6 |
| 9.90 | 9.72 | 5 | 5 |
| 10.14 | 9.90 | 6 | 4 |
| 10.28 | 10.08 | 7 | 3 |
| 10.53 | 10.28 | 8 | 2 |
| 10.83 | 10.57 | 9 | 1 |

# Ⅲ. Preparation of commonly used electrophoresis buffer

(1) Electrophoresis buffer (Table A.16)

Table A.16  Electrophoresis buffer

| Buffer | Working solution | Concentrated stock solution (per liter) |
|---|---|---|
| Tris-acetic acid(TAE) | 1×: | 50×: |
| | 0.04 mol/L Tris-acetic acid(TAE) | 242 g  Tris-base |
| | | 57.1 mL glacial acetic acid |
| | | 100 mL 0.5 mol/L EDTA(pH 8.0) |
| Tris-phosphoric acid(TPE) | 1×: | 10×: |
| | 0.09 mol/L Tris-phosphoric acid | 108 g Tris-base |
| | 0.002 mol/L EDTA | 15.5 mL 85% phosphoric acid (1.679 g/mL) |
| | | 40 mL 0.5 mol/L EDTA (pH 8.0) |
| Tris-boric acid(TBE) [a] | 0.5×: | 5×: |
| | 0.045 mol/L Tris-boric acid | 54.0 g Tris-base |
| | 0.001 mol/L EDTA | 27.5 g boric acid |
| | | 20 mL 0.5 mol/L EDTA (pH 8.0) |
| Alkaline buffer [b] | 1×: | 1×: |
| | 50 mmol/L NaOH | 5 mL 10 mol/L NaOH |
| | 1 mmol/L EDTA | 2 mL 0.5 mol/L EDTA(pH 8.0) |
| Tris-glycine | 1×: | 5×: |
| | 25 mmol/L Tris | 15.1 g Tris-base |
| | 250 mmol/L glycine | 94.0 g glycine (electrophoretic grade pH 8.3) |
| | 0.1% SDS | 50 mL 10% SDS (electrophoretic grade) |

1) The concentrated TBE solution can form deposits after long term storage. To reduce deposits, store 5×TBE solution in a regent bottle at room temperature. Once precipitation forms over time, the buffer will be discarded. Although 1×TBE (1 : 5 diluted stock solution) is usually used for agarose gel electrophoresis, 0.5×working solution contains enough buffer capacity for agarose gel electrophoresis. At present, almost all agarose gel electrophoresis uses 1 : 10 diluted stock solution as the running buffer, but for polyacrylamide gel electrophores, 1×TBE should be the running buffer. The vertical slot buffer tank of polyacrylamide gel electrophores is so small that only 1×TBE solution can supply enough buffer capacity for heavy loading effect on the circuit.

2) The alkaline electrophoretic buffer should be prepared just before use.

3) Tris-glycine buffer is used for SDS polyacrylamide gel electrophoresis.

(2) Electrophoresis loading buffers

1) 6×Loading buffer (DNA electrophoresis)

Dissolve 25 mg bromophenol blue in 6.7 mL ddH$_2$O, then add 25 mg xylene cyanide and mix fully again. Add 3.3 mL glycerol and mix fully again. Aliquot and store at −20 ℃.

2) 5×SDS-PAGE Loading buffer

Add 12.5 mL 1 mol/L Tris-HCl (pH 6.8), 5.0 g SDS, 250 mg bromophenol blue, 25 mL glycerol and 25 mL β-mercaptoethanol to a 10 mL plastic centrifuge tube, then add ddH$_2$O and mix fully, make up the final volume to 50 mL. Aliquot and store at −20 ℃.

# Ⅳ. Preparation of the common stock solution

(1) 10% ammonium peroxydisulfate(w/v)

Dissolve 1 g ammonium persulfate in 10 mL ddH$_2$O and store at 4 ℃. Ammonium persulfate decays slowly in solution, so replaces the stock solution every 2−3 weeks.

(2) 20% sodiumdodecyl sulfate (SDS) solution

Dissolve 20 g solid SDS (the molecular weight is 288.44) in 70 mL ddH$_2$O at 42 ℃, and make up the final volume to 100 mL with ddH$_2$O.

(3) 3 mol/L NaAc solution (pH 5.6)

Dissolve 49.2 g sodium acetate anhydrous in 140 mL ddH$_2$O by heating. Adjust pH value to 5.6 by adding 30 mL glacial acetic acid (the molecular weight is 60.05) and make up to 200 mL with ddH$_2$O. Sterilize 20 min and store at 4 ℃.

(4) 3 mol/L NaAc solution (pH 4.8)

Besides adjusting pH value to 4.8 by adding −40 mL glacial acetic acid, other preparation as step 3 above.

(5) 1 mol/L CaCl$_2$ solution

Dissolve 55.5 g anhydrous calcium chloride (the molecular weight is 110.99) in 300 mL ddH$_2$O. Make up the final volume to 500 mL. The solution is autoclaved at 1.1 kg/cm$^2$ for 20 min (sterilization).

(6) 1 mol/L KAc solution (pH 7.5)

Dissolve 9.82 g potassium acetate (the molecular weight is 98.14) in 90 mL ddH$_2$O. Adjust pH value to 7.5 by adding 2 mol/L acetic acid and dilute the solution with ddH$_2$O to 100 mL. Filter through 0.22 μm filter and store at room temperature.

(7) Ethidium bromide solution (EB)

1) 10 mg/mL EB solution

Dissolve 200 mg EB (the molecular weight is 394.33) in brown reagent bottle and add ddH$_2$O to make up the final volume to 20 mL. Store at 4 ℃.

2) 1 mg/mL EB solution

Pipette 10 mL of 10 mg/mL EB solution into brown reagent bottle and add 90 mL ddH$_2$O and shake slowly. Store at 4 ℃ in fridge.

*Note: ethidium bromide is the mutagenic agent of DNA and a very strong carcinogen. Wear gloves at all times.*

(8) Tris–HCl–saturated phenol solution (pH 8.0)

Most batches of commercial liquified phenol are clear and colorless and can be used in molecular cloning without redistillation. Occasionally, batches of liquified phenol are pink or yellow, and these should be rejected and returned to the manufacturer. Crystalline phenol is not recommended because it must be redistilled at 160 ℃ to remove oxidation products, such as quinones, that cause the breakdown of phosphodiester bonds or cause crosslinking of RNA and DNA.

(9) Equilibration of phenol

Before use, phenol must be equilibrated to a pH >7.8 because DNA will partition into the organic phase at acid pH.

1) Liquified phenol should be stored at −20 ℃. As needed, remove the phenol from the freezer, allow it to warm to room temperature, and then melt it at 68 ℃. Add hydroxyquinoline to a final concentration of 0.1%. This compound is an antioxidant, a partial inhibitor of RNase, and a weak chelator of metal ions. In addition, its yellow color provides a convenient way to identify the organic phase.

2) To the melted phenol, add an equal volume of buffer [usually 0.5 mol/L Tris–HCl (pH 8.0) at room temperature]. Stir the mixture on a magnetic stirrer for 15 min, and then turn off the stirrer. When the two phases have separated, aspirate as much as possible of the upper (aqueous) phase using a glass pipette attached to a vacuum line equipped with traps.

3) Add an equal volume of 0.1 mol/L Tris–HCl (pH 8.0) to the phenol. Stir the mixture on a magnetic stirrer for 15 min, and then turn off the stirrer. Remove the upper aqueous phase as described in step 2). Repeat the extractions until the pH of the phenolic phase is >7.8 (as measured with pH paper).

4) After the phenol is equilibrated and the final aqueous phase has been removed, add 0.1 volume of 0.1 mol/L Tris–HCl (pH 8.0) containing 0.2% β–mercaptoethanol. The phenol solution may be stored in this form under 100 mmol/L Tris–HCl (pH 8.0) in a light–tight bottle at 4 ℃ for periods of up to 1 month.

(10) 30% acrylamide solution (Separated protein gel mother liquid)

Dissolve 29.0 g acrylamide and 1.0 g N, N'–methylbisacrylamide in 80 mL of ddH$_2$O. Heating may be necessary to dissolve the acrylamide. Adjust the volume to 100 mL with ddH$_2$O. Sterilize the solution by filtration (0.45–micron pore size). Check pH value (should be 7.0 or less). Store in dark bottles at room temperature.

*Note: acrylamide is a neurotoxin! Use extreme care when handling solids and solutions containing acrylamide and bisacrylamide! Wear a mask and gloves when weighing out solid acrylamide.*

(11) 24 mg/mL IPTG solution

Dissolve 1.2 g IPTG in 40 mL ddH$_2$O in a 50 mL centrifuge tube, and make up the final volume to 50 mL. Filter through 0.22 μm filters. Aliquot and store at −20 ℃.

(12) 20 mg/mL X–gal solution

Weigh 1 g X–gal in a 50 mL centrifuge tube, and dissolve X–gal in 40 mL dimethylformamide (DMF) and makeup to 50 mL. Aliquot and store at −20 ℃.

# A-3  Preparation and Preservation of Bacterial Experimental Reagents

## Ⅰ. Preparation of the bacterial culture

(1) LB-medium

Dissolve 10 g tryptone, 5 g yeast extract and 10 g NaCl in -800 mL ddH$_2$O. Adjust the pH value of the medium to 7.0 by adding 5 mol/L NaOH (-0.2 mL) and make up the final volume to 1 L. Autoclave and store at 4 ℃.

(2) LB/Amp-medium

Dissolve 10 g peptone, 5 g yeast extract and 10 g NaCl in-800 mL ddH$_2$O. Adjust the pH value of the medium to 7.0 by adding 5 mol/L NaOH (-0.2 mL) and make up the final volume to 1 L. After autoclaving and cooling down to the room temperature, add 1 mL ampicillin (100 mg/mL) to the medium and mix fully. Store at 4 ℃.

(3) TB-medium

1) Phosphate buffered saline (0.17 mol/L KH$_2$PO$_4$, 0.72 mol/L K$_2$HPO$_4$) 100 mL.

Dissolve 2.31 g KH$_2$PO$_4$ and 12.54 g K$_2$HPO$_4$ in 90 mL ddH$_2$O and make up the final volume to 100 mL. Autoclave.

2) Dissolve 12 g peptone, 24 g yeast extract and 4 mL glycerol in -800 mL ddH$_2$O in 1L beaker. Mix fully and make up the final volume to 1 L.

3) After autoclaving and cooling down to -60 ℃, add 100 mL of above phosphate buffered saline and store at 4 ℃.

(4) TB/Amp-medium

1) The preparation of phosphate buffered saline (0.17 mol/L KH$_2$PO$_4$, 0.72 mol/L K$_2$HPO$_4$) 100 mL The preparation is the same as step (3)1) above.

2) Dissolve 12 g peptone, 24 g yeast extract and 4 mL glycerol in -800 mL ddH$_2$O in 1 L beaker. Mix fully and make up the final volume to 1 L.

3) After autoclaving and cooling down to 60 ℃, add 100 mL of above phosphate buffered saline and 1 mL 100 mg/mL ampicillin. Mix and store at 4 ℃.

(5) SOB-medium

1) 250 mmol/L KCl solution.

Dissolve 1.86 g KCl in 90 mL ddH$_2$O to a final volume of 100 mL. Autoclave.

2) 2 mol/L MgCl$_2$ solution.

Dissolve 19 g MgCl$_2$ in 90 mL ddH$_2$O to a final volume of 100 mL. Autoclave.

3) Dissolve 20 g peptone, 5 g yeast extract and 0.5 g NaCl in -800 mL ddH$_2$O in 1 L beaker and mix fully. Add 10 mL 250 mmol/L KCl solution to the beaker. Adjust the pH value of the medium to 7.0 with 5N NaOH (-0.2 mL) and make up the final volume to 1 L. Autoclave and store at 4 ℃. Just before use, add 5 mL sterile solution of 2 mol/L MgCl$_2$.

(6) SOC-medium

1) 1 mol/L glucose solution.

Dissolve 18 g glucose in 90 mL ddH$_2$O and mix fully. Make up the final volume to 100 mL with ddH$_2$O. Filter through 0.22 μm Filters.

2) Add 2 mL of the sterilized 1 mol/L glucose solution to 100 mL SOB-medium and mix fully. Store at

4 ℃.

(7) Common solid medium

1) Prepare the liquid medium according to the recipe of the liquid medium. Add one of the following reagents before autoclave sterilization.

Agar( agar: for agar plate)　　　　15 g/L

Agar( agar: for the top agar)　　　7 g/L

Agarose( agarose: for agarose plate)　　　15 g/L

Agarose( agarose: for the top agarose)　　　7 g/L

2) After autoclave sterilization, wear the gloves to take out the medium and stir the container to dissolve agar or agarose completely( Caution: at this moment, the culture medium is at high temperature, caution hot).

3) Cool the medium down to 50–60 ℃, add thermally unstable substance (such as an antibiotic) and stir the container completely.

4) plate medium (30–35 mL medium/90mm petri dish).

(8) LB/Amp/X–gal/IPTG plate

1) Dissolve 10 g peptone, 5 g yeast extract and 10 g NaCl in –800 mL $ddH_2O$ in 1 L beaker and mix fully.

2) Adjust pH value of the medium to 7. 0 by adding 5 mol/L NaOH solution ( –0. 2 mL) and make up the final volume to 1 L with $ddH_2O$ and then add 15 g agar.

3) After autoclave sterilization, the medium cool down to –60 ℃, add 100 mL sterilized PBS, 1 mL ampicillin (100 mg/mL), 1 mL IPTG(24 mg/mL) and 2 mL X–gal (20 mg/mL) and mix them fully. Plate medium (30–35 mL medium/90mm petri dish). Store at 4 ℃.

(9) TB/Amp/X–gal/IPTG plate culture

1) The preparation of phosphate buffer (0. 17mol/L $KH_2PO_4$, 0. 72mol/L $K_2HPO_4$) 100 mL.

The protocol is the same as the steps above.

2) Dissolve 12 g peptone, 24 g yeast extract and 4 mL glycerol in –800 mL $ddH_2O$ in 1 L beaker and mix fully, and then add 15 g agar and make up the final volume to 1 L with $ddH_2O$. After autoclave sterilization, cool down to be at 60 ℃, add 100 mL of the above phosphate buffer saline, 1 mL ampicillin (100 mg/mL), 1 mL IPTG (24 mg/mL) and 2 mL X–gal (20 mg/mL) and mix them fully. Plate medium (30–35 mL medium/90mm petri dish). Store at 4 ℃.

# Ⅱ. Preparation of antibiotic solution

The preparation of antibiotic solution are shown in Table A. 17.

Table A. 17　Preparation of antibiotic solution

| Antibiotics | Stock solution[a] (mg/mL) | Storage temperature (℃) | Working concentration (μg/mL) | | Working concentration range |
|---|---|---|---|---|---|
| | | | Stringent plasmid | Relaxed plasmid | |
| Ampicillin | 50 (solvent $H_2O$) | –20 | 20 | 60 | 25–200 |
| Carbenicillin | 50 (solvent $H_2O$) | –20 | 20 | 60 | 25–100 |

Continue to Table A. 17

| Antibiotics | Stock solution[a] (mg/mL) | Storage temperature (℃) | Working concentration (μg/mL) | | Working concentration range |
|---|---|---|---|---|---|
| | | | Stringent plasmid | Relaxed plasmid | |
| Chloramphenicol | 34 (solvent EtOH) | −20 | 25 | 170 | 10–100 |
| Kanamycin | 10 (solvent H₂O) | −20 | 10 | 50 | 25–170 |
| Streptomycin | 10 (solvent H₂O) | −20 | 10 | 50 | 10–50 |
| Tetracycline[b] | 5 (solvent EtOH) | −20 | 10 | 50 | 10–50 |

1) If water is used as a diluent to dissolve antibiotics, the working solution needs to be filtered through 0. 22 μm filters. Nevertheless, ethanol solvent can kill organisms itself including bacteria and fungi, sterilization is not required for chloramphenicol and tetracycline solution. All antibiotic stock solution should be stored in a dark container.

2) $Mg^{2+}$ is tetracycline antagonist, *E. coli* strain screened for resistance to tetracycline should be cultured in LB media without magnesium salts.

## III. Preservation methods of common Escherichia coli strains

(1) Preservation of glycerol tube

Culture 2 mL bacterial fluid overnight and take 1 mL into cryogenic vials. Add 200 μL of 80% glycerol and then turn the vials upside down. Screw tube cap tightly and seal it with parafilm. Store at −70 ℃ or −20 ℃.

(2) Plate preservation

In order to identify the different bacterial colonies inoculated on the plate, paste a circular label paper at the bottom of the new petri dish or mark some lines with a marker pen. Pick a single colony by using sterilized toothpick and inoculate it on petri dish at 37 ℃ overnight. Store at 4 ℃. Seal the edge of petri dish tightly with parafilm to protect the humidity of content in petri dish.

(3) Stab culture

Stab culture cannot preserve plasmid−carrying strains because this medium contains no antibiotics. The glass tube is filled with 2/3 volume of agar and nutrients, autoclaved and cooled down to the room temperature (the medium becomes solid). After the inoculation needle is sterilized by flaming, it needs to stab into the center of the agar to cool. Pick a single colony by the inoculation needle and then insert into the bottom of the glass tube. Culture it 2−18 h at 37 ℃. Screw the tube cap tightly and seal the edge of the bottle cap with parafilm, better with paraffin wax. Store in the dark place at room temperature.

# Appendix B

## Common Data Tables in Biochemistry Experiments

### I. Standard data of relative molecular weight of the common proteins

Table B.1　Standard data of relative molecular weight of the common proteins

| Reference standards of high-molecular-weight proteins | | Reference standards of medium-molecular-weight proteins | | Reference standards of low-molecular-weight proteins | |
|---|---|---|---|---|---|
| Myosin | 212,000 | Phosphorylase B | 97,400 | Myosin | 212,000 |
| β-galactosidase | 116,000 | Bovine serum albumin | 66,200 | β-galactosidase | 116,000 |
| Phosphorylase B | 97,400 | Ovalbumin | 43,000 | Phosphorylase B | 97,400 |
| Bovine serum albumin | 66,200 | Carbonic anhydrase | 27,600 | Bovine serum albumin | 66,200 |
| Ovalbumin | 43,000 | Soyabean trypsin inhibitor | 21,500 | Ovalbumin | 43,000 |
| | | Lysozyme | 13,750 | Carbonic anhydrase | 27,600 |
| | | | | Soybean trypsin inhibitor | 21,500 |
| | | | | Lysozyme | 13,750 |
| | | | | Trasylol | 500 |

# II . The main parameters of 20 kinds of amino acids

Table B.2    The main parameters of 20 kinds of amino acids

| Amino acids | 3-letter | 1-letter | Relative molecular weight | Isoelectric point | Polarity |
|---|---|---|---|---|---|
| Glycine | Gly | G | 75 | 5.97 | Hydrophobic |
| Alanine | Ala | A | 89 | 6.00 | Hydrophobic |
| Valine | Val | V | 117 | 5.96 | Hydrophobic |
| Leucine | Leu | L | 131 | 5.98 | Hydrophobic |
| Isoleucine | Ile | I | 131 | 6.02 | Hydrophobic |
| Methionine | Met | M | 149 | 5.74 | Hydrophobic |
| Proline | Pro | P | 115 | 6.30 | Hydrophobic |
| Phenylalanine | Phe | F | 165 | 5.48 | Hydrophobic |
| Tryptophan | Trp | W | 204 | 5.89 | Hydrophobic |
| Serine | Ser | S | 105 | 5.68 | Hydrophilic |
| Threonine | Thr | T | 119 | 5.60 | Hydrophilic |
| Asparagine | Asn | N | 133 | 5.41 | Hydrophilic |
| Glutamine | Gln | Q | 147 | 5.65 | Hydrophilic |
| Aspartic acid | Asp | D | 133 | 2.98 | Zwitterionic |
| Glutamic acid | Glu | E | 147 | 3.22 | Zwitterionic |
| Cysteine | Cys | C | 121 | 5.07 | Zwitterionic |
| Tyrosine | Tyr | Y | 181 | 5.66 | Zwitterionic |
| Histidine | His | H | 155 | 7.59 | Zwitterionic |
| Lysine | Lys | K | 146 | 9.74 | Zwitterionic |
| Arginine | Arg | R | 174 | 10.76 | Zwitterionic |

# III . The conversion data of nucleic acids and proteins

(1) The conversion of nucleic acid data

Spectrophotometric conversion

1 $A_{260}$ dsDNA = 50 μg/mL

1 $A_{260}$ ssDNA = 33 μg/mL

1 $A_{260}$ ssRNA = 40 μg/mL

(2) DNA molar conversion

1 μg 1,000 bp DNA = 1.52 pmol = 3.03 pmol terminal end

1 μg pBR322 DNA = 0.36 pmol

1 pmol 1000 bp DNA = 0.66 μg

1 pmol pBR322 = 2.8 μg

1 kb dsDNA( sodium salt) $=6.6 \times 10^5$ Da

1 kb ssDNA( sodium salt) $=3.3 \times 10^5$ Da ( the average molecular weight of dNMP$=330$ Da)

1 kb ssRNA( sodium salt) $=3.4 \times 10^5$ Da ( the average molecular weight of NMP$=345$ Da)

( 3) The conversion of protein data

Protein molar conversion

100 pmol 100,000 Da protein$=10$ μg

100 pmol 50,000 Da protein$=5$ μg

100 pmol 10,000 Da protein$=1$ μg

the average molecular weight of amino acids$=126.7$ Da

( 4) Conversion between protein and nucleic acid

Protein/DNA conversion

1 kb DNA$=333$ amino acids coding capacity $=3.7 \times 10^4$ Da protein

10,000 Da protein$=270$ bp DNA

30,000 Da protein$=810$ bp DNA

50,000 Da protein$=1.35$ kb DNA

100,000 Da protein$=2.7$ kb DNA

# Ⅳ. Standard data of relative molecular weight of the common nucleic acids

Table B.3　Standard data of relative molecular weight of the common nucleic acids

| Nucleic acid | Numbers of nucleotides | Molecular weight( Da) |
|---|---|---|
| λDNA | 48,502( circular double-stranded) | $3.0 \times 10^7$ |
| pBR322 | 4,363( double-stranded) | $2.8 \times 10^6$ |
| 28S rRNA | 4,800 | $1.6 \times 10^6$ |
| 23S rRNA | 3,700 | $1.2 \times 10^6$ |
| 18S rRNA | 1,900 | $6.1 \times 10^5$ |
| 19S rRNA | 1,700 | $5.5 \times 10^5$ |
| 5S rRNA | 120 | $3.6 \times 10^4$ |
| tRNA( *Escherichia coli*) | 5 | $2.5 \times 10^4$ |

# Ⅴ. Agarose gel concentration and optimum resolution range of linear DNA

Table B.4　Agarose gel concentration and optimum resolution range of linear DNA

| Agarose gel concentration | Optimum resolution range of linear DNA( bp) |
|---|---|
| 0.5% | 1,000–30,000 |
| 0.7% | 800–12,000 |
| 1.0% | 500–10,000 |

Continue to Table B.4

| Agarose gel concentration | Optimum resolution range of linear DNA( bp) |
|---|---|
| 1.2% | 400–7,000 |
| 1.5% | 200–3,000 |
| 2.0% | 50–2,000 |

# VI. PAGE gel recipe (for nucleic acid electrophoresis)

Table B.5    PAGE gel recipe (for nucleic acid electrophoresis)

| Gel concentration and reagents | The volume of the reagents for the different volume of PAGE gel (mL) | | | | | | | |
|---|---|---|---|---|---|---|---|---|
| | 15 mL | 20 mL | 25 mL | 30 mL | 40 mL | 50 mL | 80 mL | 100 mL |
| **3.5% gel** | | | | | | | | |
| $H_2O$ | 10.2 | 13.5 | 16.9 | 20.3 | 27.1 | 33.9 | 54.2 | 67.7 |
| 30% acrylamide | 1.7 | 2.3 | 2.9 | 3.5 | 4.6 | 5.8 | 9.3 | 11.6 |
| 10×TBE | 3.0 | 4.0 | 5.0 | 6.0 | 8.0 | 10.0 | 16.0 | 20.0 |
| 10% ammonium persulfate | 0.11 | 0.14 | 0.18 | 0.21 | 0.28 | 0.35 | 0.56 | 0.70 |
| TEMED | 0.010 | 0.013 | 0.016 | 0.020 | 0.026 | 0.033 | 0.052 | 0.065 |
| **5% gel** | | | | | | | | |
| $H_2O$ | 9.4 | 12.5 | 15.7 | 18.8 | 25.1 | 31.4 | 50.2 | 62.7 |
| 30% acrylamide | 2.5 | 3.3 | 4.2 | 5.0 | 6.6 | 8.3 | 13.3 | 16.6 |
| 10×TBE | 3.0 | 4.0 | 5.0 | 6.0 | 8.0 | 10.0 | 16.0 | 20.0 |
| 10% ammonium persulfate | 0.11 | 0.14 | 0.18 | 0.21 | 0.28 | 0.35 | 0.56 | 0.70 |
| TEMED | 0.010 | 0.013 | 0.016 | 0.020 | 0.026 | 0.033 | 0.052 | 0.065 |
| **8% gel** | | | | | | | | |
| $H_2O$ | 7.9 | 10.5 | 13.2 | 15.8 | 21.1 | 26.4 | 42.2 | 52.7 |
| 30% acrylamide | 4.0 | 5.3 | 6.7 | 8.0 | 10.6 | 13.3 | 21.3 | 26.6 |
| 10×TBE | 3.0 | 4.0 | 5.0 | 6.0 | 8.0 | 10.0 | 16.0 | 20.0 |
| 10% ammonium persulfate | 0.11 | 0.14 | 0.18 | 0.21 | 0.28 | 0.35 | 0.56 | 0.70 |
| TEMED | 0.010 | 0.013 | 0.016 | 0.020 | 0.026 | 0.033 | 0.052 | 0.065 |
| **12% gel** | | | | | | | | |
| $H_2O$ | 5.9 | 7.9 | 9.8 | 11.8 | 15.7 | 19.7 | 31.4 | 39.3 |
| 30% acrylamide | 6.0 | 8.0 | 10.0 | 12.0 | 16.0 | 20.0 | 32.0 | 40.0 |
| 10×TBE | 3.0 | 4.0 | 5.0 | 6.0 | 8.0 | 10.0 | 16.0 | 20.0 |
| 10% ammonium persulfate | 0.11 | 0.14 | 0.18 | 0.21 | 0.28 | 0.35 | 0.56 | 0.70 |
| TEMED | 0.010 | 0.013 | 0.016 | 0.020 | 0.026 | 0.033 | 0.052 | 0.065 |

Continue to Table B.5

| Gel concentration and reagents | The volume of the reagents for the different volume of PAGE gel (mL) | | | | | | | |
|---|---|---|---|---|---|---|---|---|
| | 15 mL | 20 mL | 25 mL | 30 mL | 40 mL | 50 mL | 80 mL | 100 mL |
| 20% gel | | | | | | | | |
| $H_2O$ | 1.9 | 2.5 | 3.2 | 3.8 | 5.1 | 6.4 | 10.2 | 12.7 |
| 30% acrylamide | 10.0 | 13.3 | 16.7 | 20.0 | 26.6 | 33.3 | 53.3 | 66.6 |
| 10×TBE | 3.0 | 4.0 | 5.0 | 6.0 | 8.0 | 10.0 | 16.0 | 20.0 |
| 10% ammonium persulfate | 0.11 | 0.14 | 0.18 | 0.21 | 0.28 | 0.35 | 0.56 | 0.70 |
| TEMED | 0.010 | 0.013 | 0.016 | 0.020 | 0.026 | 0.033 | 0.052 | 0.065 |

# VII. SDS–PAGE stacking gel (5% acrylamide) recipe

Table B.6　SDS–PAGE stacking gel recipe

| The reagents | The volume of the reagents for the different volume of SDS–PAGE stacking gel (mL) | | | | | | | |
|---|---|---|---|---|---|---|---|---|
| | 1 mL | 2 mL | 3 mL | 4 mL | 5 mL | 6 mL | 8 mL | 10 mL |
| $H_2O$ | 0.68 | 1.40 | 2.10 | 2.70 | 3.40 | 4.10 | 5.50 | 6.80 |
| 30% acrylamide | 0.17 | 0.33 | 0.50 | 0.67 | 0.83 | 1.00 | 1.30 | 1.7 |
| 1.0 mol/L Tris (pH 6.8) | 0.13 | 0.25 | 0.38 | 0.50 | 0.63 | 0.75 | 1.00 | 1.25 |
| 10% SDS | 0.01 | 0.02 | 0.03 | 0.04 | 0.05 | 0.06 | 0.08 | 0.10 |
| 10% ammonium persulfate | 0.01 | 0.02 | 0.03 | 0.04 | 0.05 | 0.06 | 0.08 | 0.10 |
| TEMED | 0.001 | 0.002 | 0.003 | 0.004 | 0.005 | 0.006 | 0.008 | 0.010 |

# VIII. SDS–PAGE resolving gel recipe

Table B.7　SDS–PAGE resolving gel recipe

| The reagents | The volume of the reagents for the different volume of SDS–PAGE resolving gel (mL) | | | | | | | |
|---|---|---|---|---|---|---|---|---|
| | 5 mL | 10 mL | 15 mL | 20 mL | 25 mL | 30 mL | 40 mL | 50 mL |
| 6% gel | | | | | | | | |
| $H_2O$ | 2.6 | 5.3 | 7.9 | 10.6 | 13.2 | 15.9 | 21.2 | 26.5 |
| 30% acrylamide | 1.0 | 2.0 | 3.0 | 4.0 | 5.0 | 6.0 | 8.0 | 10.0 |
| 1.5 mol/L Tris–HCl (pH 8.8) | 1.3 | 2.5 | 3.8 | 5.0 | 6.3 | 7.5 | 10.0 | 12.5 |
| 10% SDS | 0.05 | 0.1 | 0.15 | 0.2 | 0.25 | 0.3 | 0.4 | 0.5 |
| 10% ammonium persulfate | 0.05 | 0.1 | 0.15 | 0.2 | 0.25 | 0.3 | 0.4 | 0.5 |
| TEMED | 0.004 | 0.008 | 0.012 | 0.016 | 0.020 | 0.024 | 0.032 | 0.040 |

Continue to Table B.7

| The reagents | The volume of the reagents for the different volume of SDS–PAGE resolving gel (mL) | | | | | | | |
|---|---|---|---|---|---|---|---|---|
| | 5 mL | 10 mL | 15 mL | 20 mL | 25 mL | 30 mL | 40 mL | 50 mL |
| **8% gel** | | | | | | | | |
| H$_2$O | 2.3 | 4.6 | 6.9 | 9.3 | 11.5 | 13.9 | 18.5 | 23.2 |
| 30% acrylamide | 1.3 | 2.7 | 4.0 | 5.3 | 6.7 | 8.0 | 10.7 | 13.3 |
| 1.5 mol/L Tris–HCl (pH 8.8) | 1.3 | 2.5 | 3.8 | 5.0 | 6.3 | 7.5 | 10.0 | 12.5 |
| 10% SDS | 0.05 | 0.1 | 0.15 | 0.2 | 0.25 | 0.3 | 0.4 | 0.5 |
| 10% ammonium persulfate | 0.05 | 0.1 | 0.15 | 0.2 | 0.25 | 0.3 | 0.4 | 0.5 |
| TEMED | 0.003 | 0.006 | 0.009 | 0.012 | 0.015 | 0.018 | 0.024 | 0.030 |
| **10% gel** | | | | | | | | |
| H$_2$O | 1.9 | 4.0 | 5.9 | 7.9 | 9.9 | 11.9 | 15.9 | 19.8 |
| 30% acrylamide | 1.7 | 3.3 | 5.0 | 6.7 | 8.3 | 10.0 | 13.3 | 16.7 |
| 1.5 mol/L Tris–HCl (pH 8.8) | 1.3 | 2.5 | 3.8 | 5.0 | 6.3 | 7.5 | 10.0 | 12.5 |
| 10% SDS | 0.05 | 0.1 | 0.15 | 0.2 | 0.25 | 0.3 | 0.4 | 0.5 |
| 10% ammonium persulfate | 0.05 | 0.1 | 0.15 | 0.2 | 0.25 | 0.3 | 0.4 | 0.5 |
| TEMED | 0.002 | 0.004 | 0.006 | 0.008 | 0.010 | 0.012 | 0.016 | 0.020 |
| **12% gel** | | | | | | | | |
| H$_2$O | 1.6 | 3.3 | 4.9 | 6.6 | 8.2 | 9.9 | 13.2 | 16.5 |
| 30% acrylamide | 2.0 | 4.0 | 6.0 | 8.0 | 10.0 | 12.0 | 16.0 | 20.0 |
| 1.5 mol/L Tris–HCl (pH 8.8) | 1.3 | 2.5 | 3.8 | 5.0 | 6.3 | 7.5 | 10.0 | 12.5 |
| 10% SDS | 0.05 | 0.1 | 0.15 | 0.2 | 0.25 | 0.3 | 0.4 | 0.5 |
| 10% ammonium persulfate | 0.05 | 0.1 | 0.15 | 0.2 | 0.25 | 0.3 | 0.4 | 0.5 |
| TEMED | 0.002 | 0.004 | 0.006 | 0.008 | 0.010 | 0.012 | 0.016 | 0.020 |
| **15% gel** | | | | | | | | |
| H$_2$O | 1.1 | 2.3 | 3.4 | 4.6 | 5.7 | 6.9 | 9.2 | 11.5 |
| 30% acrylamide | 2.5 | 5.0 | 7.5 | 10 | 12.5 | 15.0 | 20.0 | 25.0 |
| 1.5 mol/L Tris–HCl (pH 8.8) | 1.3 | 2.5 | 3.8 | 5.0 | 6.3 | 7.5 | 10.0 | 12.5 |
| 10% SDS | 0.05 | 0.1 | 0.15 | 0.2 | 0.25 | 0.3 | 0.4 | 0.5 |

Continue to Table B.7

| The reagents | The volume of the reagents for the different volume of SDS–PAGE resolving gel ( mL) | | | | | | | |
|---|---|---|---|---|---|---|---|---|
| | 5 mL | 10 mL | 15 mL | 20 mL | 25 mL | 30 mL | 40 mL | 50 mL |
| 10% ammonium persulfate | 0.05 | 0.1 | 0.15 | 0.2 | 0.25 | 0.3 | 0.4 | 0.5 |
| TEMED | 0.002 | 0.004 | 0.006 | 0.008 | 0.010 | 0.012 | 0.016 | 0.020 |

# IX. The concentration and the best separation range for SDS–PAGE resolving gel

Table B.8    The concentration and the best separation range for SDS–PAGE resolving gel

| The concentration of SDS–PAGE resolving gel | The best separation range of the protein ( kD) |
|---|---|
| 6% | 50–150 |
| 8% | 30–90 |
| 10% | 20–80 |
| 12% | 12–60 |
| 15% | 10–40 |

*Zhou Hongbo*